影印版说明

MOMENTUM PRESS 出版的 *Plastics Technology Handbook*（2 卷）是介绍塑料知识与技术的大型综合性手册，内容涵盖了从高分子基本原理，到塑料的合成、种类、性能、配料、加工、制品，以及模具、二次加工等各个方面。通过阅读、学习本手册，无论是专业人员还是非专业人员，都会很快熟悉和掌握塑料制品的设计和制造方法。可以说一册在手，别无他求。

原版 2 卷影印时分为 11 册，第 1 卷分为：

    塑料基础知识·塑料性能
    塑料制品生产
    注射成型
    挤压成型
    吹塑成型
    热成型
    发泡成型·压延成型

第 2 卷分为：

    涂层·浇注成型·反应注射成型·旋转成型
    压缩成型·增强塑料·其他工艺
    模具
    辅机与二次加工设备

唐纳德 V·罗萨多，波士顿大学化学学士学位，美国东北大学 MBA 学位，马萨诸塞大学洛厄尔分校工程塑料和加州大学工商管理博士学位（伯克利）。著有诸多论文及著作，包括《塑料简明百科全书》、《注塑手册（第三版）》以及塑料产品材料和工艺选择手册等。活跃于塑料界几十年，现任著名的 Plasti Source Inc. 公司总裁，并是美国塑料工业协会（SPI）、美国塑料学会（PIA）和 SAMPE（The Society for the Advancement of Material and Process Engineering）的重要成员。

    材料科学与工程图书工作室
    联系电话 0451-86412421
            0451-86414559
    邮   箱 yh_bj@aliyun.com
           xuyaying81823@gmail.com
           zhxh6414559@aliyun.com

影印版

PLASTICS TECHNOLOGY HANDBOOK

塑料技术手册

VOLUME 1

BLOW MOLDING

吹塑成型

EDITED BY

DONALD V. ROSATO

MARLENE G. ROSATO

NICK R. SCHOTT

哈爾濱工業大學出版社

HITP HARBIN INSTITUTE OF TECHNOLOGY PRESS

黑版贸审字08-2014-092号

Donald V.Rosato, Marlene G.Rosato, Nick R.Schott
Plastics Technology Handbook Volume 1
9781606500798
Copyright © 2010 by Momentum Press, LLC
All rights reserved.

Originally published by Momentum Press, LLC
English reprint rights arranged with Momentum Press, LLC through McGraw-Hill Education (Asia)

This edition is authorized for sale in the People's Republic of China only, excluding Hong Kong, Macao SAR and Taiwan.

本书封面贴有McGraw-Hill Education公司防伪标签，无标签者不得销售。
版权所有，侵权必究。

图书在版编目（CIP）数据

塑料技术手册. 第1卷. 吹塑成型 =Plastics technology handbook volume 1 blow molding：英文/（美）罗萨多（Rosato, D. V.）等主编. —影印本. — 哈尔滨：哈尔滨工业大学出版社, 2015.6
　　ISBN 978-7-5603-5044-8

Ⅰ.①塑… Ⅱ.①罗… Ⅲ.①塑料–技术手册–英文②塑料成型–吹塑–技术手册–英文 Ⅳ.①TQ320.6-62

中国版本图书馆CIP数据核字（2014）第280071号

责任编辑	许雅莹　张秀华　杨　桦
出版发行	哈尔滨工业大学出版社
社　　址	哈尔滨市南岗区复华四道街10号 邮编150006
传　　真	0451-86414749
网　　址	http://hitpress.hit.edu.cn
印　　刷	哈尔滨市石桥印务有限公司
开　　本	787mm×960mm　1/16　印张 10
版　　次	2015年6月第1版　2015年6月第1次印刷
书　　号	ISBN 978-7-5603-5044-8
定　　价	50.00元

（如因印刷质量问题影响阅读，我社负责调换）

Plastics Technology Handbook

VOLUME 1

EDITED BY

Donald V. Rosato, PhD, MBA, MS, BS, PE
Plastisource Inc.
Society of Plastics Engineers
Plastics Pioneers Association
University of Massachusetts–Lowell Plastics Advisory Board

Marlene G. Rosato, BASc (ChE) P Eng
Gander International Inc.
Canadian Society of Chemical Engineers
Association of Professional Engineers of Ontario
Product Development and Management Association

Nick R. Schott, PhD, MS, BS (ChE), PE
University of Massachusetts–Lowell Professor of Plastics Engineering & Plastics Department Head (Retired)
Plastics Institute of America
Secretary & Director for Educational and Research Programs

Momentum Press, LLC, New York

Contents

LIST OF FIGURES	9
LIST OF TABLES	15
PREFACE	17
ABOUT THE EDITORS	20
6. BLOW MOLDING	**1005**
INTRODUCTION	1005
Container	1009
Industry Size	1015
BLOW MOLDING PROCESS	1016
Blowing Requirements	1016
Airflow Control	1017
Extrusion versus Injection Blow Molding	1021
BASICS IN PROCESSING	1021
EXTRUSION BLOW MOLDING	1022
Extruder	1022
Melt Flow	1023
Parison Sag	1029
Parison Head	1034
Parison Wall Thickness	1035
Machine Design	1039
Single-Stage Design	1043
Two-Stage Design	1043
Continuous Extrusion Design	1044
Intermittent Extrusion Design	1046
INJECTION BLOW MOLDING	1063

STRETCH BLOW MOLDING — 1071
- Injection Stretch Blow Molding — 1072
- Special Machines — 1084
- Extrusion Stretch Blow Molding — 1084
- Dip Blow Molding — 1085
- Multibloc Blow Molding — 1086
- Other Blow-Molding Processes — 1086
- Blow Molding with Rotation — 1095

MOLD — 1097
- Basic Features — 1100
- Materials of Construction — 1101
- Pinch-Off Zone — 1101
- Flash Control — 1105
- Blowing and Calibrating Device — 1107
- Venting and Surface Finish — 1107
- Cooling — 1108

PLASTIC MATERIAL — 1113
- Blow Molding and Plastic — 1120
- Behavior of Plastics — 1123
- Barrier Plastic — 1125
- Barrier Material Type — 1130
- Blow Molding Reinforced Plastic — 1130

DESIGN — 1131
- Bottle Design — 1132
- Industrial Products — 1132
- Complex Irregular Shape — 1133
- Oriented 3-D Parison — 1135
- Other Design Approaches — 1136

SUMMARY — 1136
- History — 1136

Figures

Figure 6.1	Examples of extrusion, injection, and stretch blow-molding techniques	1006
Figure 6.2	Examples of the different forms of blow molding	1006
Figure 6.3	Montage of commercial and industrial blow-molded products	1007
Figure 6.4	Examples of blow-molded foodstuff containers	1008
Figure 6.5	Example of longneck blow-molded products	1008
Figure 6.6	Blow-molded containers for potato chips	1009
Figure 6.7	Examples of two sizes of blow-molded containers	1009
Figure 6.8	Blow-molded ribbed-panel automotive floor	1010
Figure 6.9	Complex 3-D blow-molded products	1010
Figure 6.10	Plastic blow-molded fuel tank (left) compared to a metal fuel tank	1011
Figure 6.11	Blow-molded aerodynamic truck wind spoiler	1012
Figure 6.12	Blow-molded 52-gallon hot-water heater that is jacketed by filament winding (chapter 15) to meet UL burst strength requirements	1012
Figure 6.13	Blow-molded water flotation wheels	1013
Figure 6.14	Blow-molded swimming pool (courtesy of Vogue Pool Products, La Salle, Quebec, Canada)	1013
Figure 6.15	Blow-molded bellow boots for automotive and other markets	1014
Figure 6.16	Sequential extruded blow-molded polypropylene automotive air duct	1014
Figure 6.17	Three locations for air to enter extrusion blow molds	1018
Figure 6.18	Blow-molding pin with escape channel for the blown air	1020

Figure 6.19	Basic processing steps in extrusion blow molding: (a) extruded heated plastic parison, mold open; (b) mold closed and bottle blown; and (c) finished bottle removed from mold	1022
Figure 6.20	Schematic of extrusion blow molding a single parison	1023
Figure 6.21	Schematic of the plastic melting action in an extruder that has two exiting parisons	1024
Figure 6.22	Relating thicknesses of swell ratio of parison and BM product	1027
Figure 6.23	Problems encountered in "countering" high-weight swell	1028
Figure 6.24	Effect of land length on swell	1029
Figure 6.25	Parison length vs. time curves for three different situations	1031
Figure 6.26	Oscillating melt flow rate near slip discontinuity of flow curve	1032
Figure 6.27	Simplified view of a heart-shaped parison die head	1034
Figure 6.28	Details of a heart-shaped parison die head	1035
Figure 6.29	Side view of center-fed die with spider supports for its core; top view: examples of four-spider support system or use of a perforated screen	1036
Figure 6.30	Examples of a grooved-core parison die head	1037
Figure 6.31	Example of double-sided parison feedhead so that a double-layered parison is produced that overlaps weld lines 180° apart (courtesy of Graham Machinery Group)	1038
Figure 6.32	Explanations of a parison die head	1039
Figure 6.33	Examples of parison wall thickness control by axial movement of the mandrel	1040
Figure 6.34	Examples of convergent and divergent die-head tooling	1040
Figure 6.35	Examples of programmed parisons	1041
Figure 6.36	Example of rectangular parison shapes where (a) die opening had a uniform thickness resulting in weak corners and (b) die opening was designed to meet the thickness requirements required	1042
Figure 6.37	Simplified schematic showing parts of a blow-molding machine	1042
Figure 6.38	Examples of preparing cut-to-size parisons for a two-stage extrusion blow-molding process (courtesy of SIG Plastics International)	1043
Figure 6.39	Introduction to a continuous extruded blow-molding system with its accumulator die head	1044
Figure 6.40	Examples of continuous extruded blow-molding systems with calibrated necks	1045
Figure 6.41	Schematics of continuous two-mold and multimold shuttle systems	1046
Figure 6.42	View of a three-milk bottle mold shuttle system	1046

Figure 6.43	Schematic of dual-sided shuttle with six parisons (courtesy of Graham Machinery Group)	1047
Figure 6.44	Closeup of dual-sided shuttle with six parisons (courtesy of Graham Machinery Group)	1048
Figure 6.45	Dual-sided shuttle with six parisons with safety doors opened (courtesy of Graham Machinery Group)	1049
Figure 6.46	Dual-sided shuttle with six parisons with safety doors closed (courtesy of Graham Machinery Group)	1049
Figure 6.47	Overcoming shuttle machine limitations (courtesy of Graham Machinery Group)	1050–1052
Figure 6.48	Schematics of continuous horizontal or vertical wheel machines	1053
Figure 6.49	Schematics of vertical wheel machine in a production line (courtesy of Graham Machinery Group)	1053
Figure 6.50	Rotary machine with closeup of rotary wheel (courtesy of Graham Machinery Group)	1054
Figure 6.51	Schematic side view of five-station rotary wheel (courtesy of Graham Machinery Group)	1055
Figure 6.52	Rotary shuttle advantages (courtesy of Graham Machinery Group)	1056–1060
Figure 6.53	Example of a reciprocating screw intermittent extrusion blow-molding machine	1061
Figure 6.54	Series of conventional horizontal injection-molding machines with appropriate blow-molding dies	1062
Figure 6.55	Example of an intermittent accumulator head extrusion blow-molding machine	1062
Figure 6.56	Example of an intermittent ram-accumulator extrusion blow-molding machine	1063
Figure 6.57	Example of the extrusion blow-molding cycle with an accumulator	1063
Figure 6.58	Schematic of an assembled intermittent accumulator parison head (courtesy of Graham Machinery Group)	1064–1065
Figure 6.59	Example of intermittent accumulator parison head (courtesy of Bekum)	1066
Figure 6.60	Example of intermittent accumulator parison head with a calibrated neck finish	1066
Figure 6.61	Example of intermittent accumulator parison head with overflow melts in the parison to eliminate weld lines	1067
Figure 6.62	Schematic of an EBM with an intermittent accumulator that	

	is fully automatic; insert is an example of a 20-liter (5-gallon) PC plastic bottle fabricated in this machine (courtesy of SIG Blowtec 2-20/30 of SIG Plastics)	1068
Figure 6.63	Intermittent extrusion blow-molding machine with accumulator molding large tanks (courtesy of Graham Machinery Group)	1069
Figure 6.64	Left view shows an injection-molded preform designed to obtain a uniform wall thickness when blow molded (right view)	1070
Figure 6.65	Example of the injection blow-molding cycle	1070
Figure 6.66	Three-station injection blow-molding system	1071
Figure 6.67	Example of ejecting blown containers using a stripper plate	1072
Figure 6.68	Examples of three-station and four-station injection blow-molding machines	1073
Figure 6.69	View of a shuttle mold to fabricate injection-molded containers	1074
Figure 6.70	Schematic of injection blow mold with a solid handle	1075
Figure 6.71	Simple handles (ring, strap, etc.) can be molded with blow-molded bottles and other products	1075
Figure 6.72	Single-stage injection stretch-blow process	1076
Figure 6.73	Schematic of the steps taken for injection stretch blow molding	1076
Figure 6.74	Schematic and internal view of a fast-operating reheat preform for stretched IBM (courtesy of SIG Plastics International)	1077
Figure 6.75	Easy-to-operate and control in-line stretch IBM (courtesy of Milacron)	1078
Figure 6.76	Example of a single-stage injection stretch blow-molding production line	1079
Figure 6.77	Temperature range for stretch blow molding polypropylene	1080
Figure 6.78	Example of stretched injection blow molding using a rod	1080
Figure 6.79	Example of stretched injection blow molding by gripping and stretching the preform	1081
Figure 6.80	Schematic of a two-step injection stretch blow-molding process (courtesy of Milacron)	1081
Figure 6.81	Example of a bottling plant using the two-step injection stretch blow-molding process	1082
Figure 6.82	Example of a two-stage injection stretch blow-molding production line	1083
Figure 6.83	Stages in the dip blow-molding process	1085
Figure 6.84	Multibloc blow-molding process	1086
Figure 6.85	Example of a six-layer coextruded blow-molded bottle	1087

Figure 6.86	Example of a five-layer coinjection blow-molded bottle	1088
Figure 6.87	Example of a five-layer coinjection blow-molded ketchup bottle	1088
Figure 6.88	Example of a three-layer coextrusion parison blow-molded head with die profiling	1089
Figure 6.89	Example of a five-layer coextrusion parison blow-molded head with die profiling (courtesy of Graham Machinery Group)	1090
Figure 6.90	Example of hot-filling PET bottle at 80° to 95°C (courtesy of SIG Plastics International)	1091
Figure 6.91	Examples of different shaped sequential extrusion blow-molding products	1093
Figure 6.92	Example of container-filling steps in the blow/fill/seal extrusion blow-molding process	1094
Figure 6.93	Example of a 3-D extrusion blow molding process (courtesy of Placo)	1094
Figure 6.94	Examples of multiple side action 3-D extrusion blow-molding molds	1095
Figure 6.95	Example of six-axis robotic control to manipulate a parison in a 3-D mold cavity to extrusion blow mold products (courtesy of SIG Plastics International)	1096
Figure 6.96	Example of a suction 3-D extrusion blow-molding process (courtesy of SIG Plastics International)	1097
Figure 6.97	Example of sequential 3-D coextrusion blow-molding machine (courtesy of SIG Plastics International)	1098
Figure 6.98	Examples of 3-D extrusion blow-molded products in their mold cavities (courtesy of SIG Plastics International)	1099
Figure 6.99	Schematic for molding with rotation using a two-stage blow-molding procedure	1099
Figure 6.100	Example of an extrusion blow mold	1101
Figure 6.101	Blow-molded corrugated bellow part between its mold halves	1102
Figure 6.102	Examples of parting line locations and other parts of a mold	1103
Figure 6.103	Example of a three-part mold to fabricate a complex threaded lid	1104
Figure 6.104	Examples of pinch-off zones in an extrusion blow mold	1105
Figure 6.105	Examples of pinch-off designs to meet requirements for different plastics and contours	1106
Figure 6.106	Example of a trapezoidal cross-section insert at the parting line	1107
Figure 6.107	Example of a calibrating blow pin	1108
Figure 6.108	Example of blow needle	1109
Figure 6.109	Example of air vent slots in an injection molding of a preform mold	1110

Figure 6.110	View of a multicavity preform mold in the background with blow molds and molded bottles in front (courtesy of SIG Plastics International)	1110
Figure 6.111	Examples of water flood cooling blow-molding molds	1113
Figure 6.112	Examples of effects of the blow-molding extruder and plastic variables on product performances	1122
Figure 6.113	Nomogram for injection blow-molded preform shot weight, cycle time, and resin use	1123
Figure 6.114	Comonomer concentrations vs. barrier properties of crystalline structures	1129
Figure 6.115	Examples of extruded blow-molded double-wall HDPE carrying case, which protects and simplifies part storage	1134
Figure 6.116	A shuttle EBM machine limitation and solution (courtesy of	
Figure 6.117	Views of multiple action extrusion blow-molding containers	1138
Figure 6.118	Schematics of moving molds and removing bottleneck flash (courtesy of Uniloy Milacron)	1138
Figure 6.119	Example of inserting a plastic injection-molded reinforcement into a blow mold	1139
Figure 6.120	Living hinge is part of the extruded blow-molding parison	1139
Figure 6.121	Collapsible bottle capable of 85% size reduction or 75% volume reduction	1139

TABLES

Table 6.1	Examples of extrusion vs. injection blow-molding performances	1016
Table 6.2	Examples of air blowing pressure required for certain plastics	1017
Table 6.3	Guide to air entrance orifice size	1019
Table 6.4	Discharge cu ft/s @ 14.7 psi and 70°F with extrusion blow time formula	1020
Table 6.5	Example of temperature conditions in an extruder plasticator based on processing different plastics	1024
Table 6.6	Examples of extruder output rates based on processing HDPE	1025
Table 6.7	Examples of plastic melt parison swell	1027
Table 6.8	General effect of shear rate on die swell of various thermoplastics	1030
Table 6.9	Examples of plastic melt and stretch temperatures	1075
Table 6.10	Examples of stretch ratios for different plastics	1084
Table 6.11	Mold design checklist	1100
Table 6.12	Examples of materials used in the construction of blow-molding molds	1104
Table 6.13	Cooling characteristics	1111
Table 6.14	Cooling temperature requirements	1111
Table 6.15	Examples of blow-molding mold cavity temperatures based on plastic being processed	1112
Table 6.16	Examples of computer software information generated and typical problems it can solve (chapter 25)	1112
Table 6.17	Examples of properties of thermoplastic bottles	1114–1115

Table 6.18	Examples of various plastics suitable for plastic liquor bottles	1116
Table 6.19	Important properties of extrusion blow-molded products and the desired goal(s) for each	1116
Table 6.20	Changes in extrusion blow-molded bottle properties resulting from resin properties	1117
Table 6.21	Changes in extrusion bold-molded blow properties resulting from changes in extrusion and molding conditions	1118
Table 6.22	Gas barrier transmission comparisons for a 24 fl oz (689 cm^3) container weighing 40 g	1119
Table 6.23	Volume shrinkage of stretch blow-molded bottles	1119
Table 6.24	Tensile test data of PET plastic	1119
Table 6.25	Guide to plastics processing temperatures for blow molding	1120
Table 6.26	Examples of fabricating conditions on blow-molded PE bottles	1121
Table 6.27	EVOH plastic range of properties	1129
Table 6.28	Examples of barrier properties of commercially available plastics	1130

Preface

This book, as a four-volume set, offers a simplified, practical, and innovative approach to understanding the design and manufacture of products in the world of plastics. Its unique review will expand and enhance your knowledge of plastic technology by defining and focusing on past, current, and future technical trends. Plastics behavior is presented to enhance one's capability when fabricating products to meet performance requirements, reduce costs, and generally be profitable. Important aspects are also presented for example to gain understanding of the advantages of different materials and product shapes. Information provided is concise and comprehensive.

Prepared with the plastics technologist in mind, this book will be useful to many others. The practical and scientific information contained in this book is of value to both the novice including trainees and students, and the most experienced fabricators, designers, and engineering personnel wishing to extend their knowledge and capability in plastics manufacturing including related parameters that influence the behavior and characteristics of plastics. The tool maker (mold, die, etc.), fabricator, designer, plant manager, material supplier, equipment supplier, testing and quality control personnel, cost estimator, accountant, sales and marketing personnel, new venture type, buyer, vendor, educator/trainer, workshop leader, librarian, industry information provider, lawyer, and consultant can all benefit from this book. The intent is to provide a review of the many aspects of plastics that range from the elementary to practical to the advanced and more theoretical approaches. People with different interests can focus on and interrelate across subjects in order to expand their knowledge within the world of plastics.

Over 20000 subjects covering useful pertinent information are reviewed in different chapters contained in the four volumes of this book, as summarized in the expanded table of contents and index. Subjects include reviews on materials, processes, product designs, and so on. From a pragmatic standpoint, any theoretical aspect that is presented has been prepared so that the practical person will understand it and put it to use. The theorist, in turn will gain an insight into the practical limitations that exist in plastics as they exist in other materials such as steel, wood, and so on. There is no material that is "perfect." The four volumes of this book together contain 1800 plus figures and 1400 plus tables providing extensive details to supplement the different subjects.

In working with any material (plastics, metal, wood, etc.), it is important to know its behavior in order to maximize product performance relative to cost/efficiency. Examples of different plastic materials and associated products are reviewed with their behavior patterns. Applications span toys, medical devices, cars, boats, underwater devices, containers, springs, pipes, buildings, aircraft, and spacecraft. The reader's product to be designed and/or fabricated can directly or indirectly be related to products reviewed in this book. Important are behaviors associated with and interrelated with the many different plastics materials (thermoplastics, thermosets, elastomers, reinforced plastics) and the many fabricating processes (extrusion, injection molding, blow molding, forming, foaming, reaction injection molding, and rotational molding). They are presented so that the technical or nontechnical reader can readily understand the interrelationships of materials to processes.

This book has been prepared with the awareness that its usefulness will depend on its simplicity and its ability to provide essential information. An endless amount of data exists worldwide for the many plastic materials that total about 35000 different types. Unfortunately, as with other materials, a single plastic material does not exist that will meet all performance requirements. However, more so than with any other materials, there is a plastic that can be used to meet practically any product requirement(s). Examples are provided of different plastic products relative to critical factors ranging from meeting performance requirements in different environments to reducing costs and targeting for zero defects. These reviews span small to large and simple to complex shaped products. The data included provide examples that span what is commercially available. For instance, static physical properties (tensile, flexural, etc.), dynamic physical properties (creep, fatigue, impact, etc.), chemical properties, and so on, can range from near zero to extremely high values, with some having the highest of any material. These plastics can be applied in different environments ranging from below and on the earth's surface, to outer space.

Pitfalls to be avoided are reviewed in this book. When qualified people recognize the potential problems that can exist, these problems can be designed around or eliminated so that they do not affect the product's performance. In this way, costly pitfalls that result in poor product performance or failure can be reduced or eliminated. Potential problems or failures are reviewed with solutions also presented. This failure/solution review will enhance the intuitive skills of people new to plastics as well as those who are already working in plastics. Plastic materials have been produced worldwide over many years for use in the design and fabrication of all kinds of plastic products that profitably and successfully meet high quality, consistency, and long-life standards. All that is needed is to understand the behavior of plastics and properly apply these behaviors.

Patents or trademarks may cover certain of the materials, products, or processes presented. They are discussed for information purposes only and no authorization to use these patents or trademarks is given or implied. Likewise, the use of general descriptive names, proprietary names, trade names, commercial designations, and so on does not in any way imply that they may be used freely. While the information presented represents useful information that can be studied or analyzed and is believed to be true and accurate, neither the authors, contributors, reviewers, nor the publisher can accept any legal responsibility for any errors, omissions, inaccuracies, or other factors. Information is provided without warranty of any kind. No representation as to accuracy, usability, or results should be inferred.

Preparation for this book drew on information from participating industry personnel, global industry and trade associations, and the authors' worldwide personal, industrial, and teaching experiences.

DON & MARLENE ROSATO AND NICK SCHOTT, 2010

About the Editors

Dr. Donald V. Rosato, president of PlastiSource, Inc., a prototype manufacturing, technology development, and marketing advisory firm in Massachusetts, United States, is internationally recognized as a leader in plastics technology, business, and marketing. He has extensive technical, marketing, and plastics industry business experience ranging from laboratory testing to production to marketing, having worked for Northrop Grumman, Owens-Illinois, DuPont/Conoco, Hoechst Celanese/Ticona, and Borg Warner/G.E. Plastics. He has developed numerous polymer-related patents and is a participating member of many trade and industry groups. Relying on his unrivaled knowledge of the industry plus high-level international contacts, Dr. Rosato is also uniquely positioned to provide an expert, inside view of a range of advanced plastics materials, processes, and applications through a series of seminars and webinars. Among his many accolades, Dr. Rosato has been named Engineer of the Year by the Society of Plastics Engineers. Dr. Rosato has written extensively, authoring or editing numerous papers, including articles published in the *Encyclopedia of Polymer Science and Engineering*, and major books, including the *Concise Encyclopedia of Plastics*, *Injection Molding Handbook 3rd ed.*, *Plastic Product Material and Process Selection Handbook*, *Designing with Plastics and Advanced Composites*, and *Plastics Institute of America Plastics Engineering, Manufacturing and Data Handbook*. Dr. Rosato holds a BS in chemistry from Boston College, MBA at Northeastern University, MS in plastics engineering from University of Massachusetts Lowell, and PhD in business administration at University of California, Berkeley.

Marlene G. Rosato, with stints in France, China, and South Korea, has very comprehensive international plastics and elastomer business experience in technical support, plant start-up and troubleshooting, manufacturing and engineering management, business development and strategic planning with Bayer/Polysar and DuPont and does extensive international technical, manufacturing, and management consulting as president of Gander International Inc. She also has an extensive writing background authoring or editing numerous papers and major books, including the *Concise Encyclopedia of Plastics*, *Injection Molding Handbook 3rd ed.*, and the *Plastics Institute of America Plastics Engineering, Manufacturing and Data Handbook*. A senior member of the Canadian Society of Chemical Engineering and the Association of Professional Engineers of Canada, Ms. Rosato is a licensed professional engineer of Ontario, Canada. She received a Bachelor of Applied Science in chemical engineering from the University of British Columbia with continuing education at McGill University in Quebec, Queens University and the University of Western Ontario both in Ontario, Canada, and also has extensive executive management training.

Professor Nick Schott, a long-time member of the world-renowned University of Massachusetts Lowell Plastics Engineering Department faculty, served as its department head for a quarter of a century. Additionally, he founded the Institute for Plastics Innovation, a research consortium affiliated with the university that conducts research related to plastics manufacturing, with a current emphasis on bioplastics, and served as its director from 1989 to 1994. Dr. Schott has received numerous plastics industry accolades from the SPE, SPI, PPA, PIA, as well as other global industry associations and is renowned for the depth of his plastics technology experience, particularly in processing-related areas. Moreover, he is a quite prolific and requested industry presenter, author, patent holder, and product/process developer, in addition to his quite extensive and continuing academic responsibilities at the undergraduate to postdoctoral level. Among America's internationally recognized plastics professors, Dr. Nick R. Schott most certainly heads everyone's list not only within the 2500 plus global UMASS Lowell Plastics Engineering alumni family, which he has helped grow, but also in broad global plastics and industrial circles. Professor Schott holds a BS in ChE from UC Berkeley, and an MS and PhD from the University of Arizona.

CHAPTER 6
BLOW MOLDING

INTRODUCTION

Blow molding (BM) is a process for converting thermoplastics into hollow objects. Like injection molding, the process is discontinuous or batchwise in nature, involving a sequence of operations that culminates in the production of a molding. This sequence or cycle is repeated automatically or semiautomatically to produce a stream of molded parts (3, 9, 23, 109, 221, 456). The three basic processes are shown in Figure 6.1. Blow molding is a very highly developed process with variant forms as summarized in Figure 6.2.

An advantage in blow molding is that of manufacturing molded products economically, in unlimited quantities, with virtually no finishing or secondary equipment required for most products. The basic process is to inflate a softened thermoplastic hollow form against the cooled surface of a closed mold where the plastic solidifies into a hollow product. The surfaces of the moldings are smooth and bright or as grained and engraved as the surfaces of the mold cavities in which they are processed.

With plastics in blow-molded products, to a greater extent than other materials, an opportunity will always exist to optimize their use since new and useful developments in materials and processing continually are on the horizon. Blow molding permits potential of consolidation of parts.

Selecting the right plastic requires applying certain factors such as, setting up performance target requirements, choosing and adapting the process to be used, and intelligently preparing a specification purchase document and work order (Fig. 1.37). Recognize that with the many varying properties of the different plastics, there are those that meet high performance requirements, such as long-time creep resistance, fatigue endurance, toughness, and so on (chapter 19). Conversely, there are those plastics that are volume- and cost-driven in their uses (chapter 30).

Applications for blow moldings are used to contain many different products, including foodstuffs, beverages, household products (appliances, air-conditioners, furniture at homes/offices/hospitals/sports arenas, etc.), personal care products, medicines/pharmaceutical products, automotive parts (bumpers, spoilers, seat backs, etc.), construction panels, tote-boxes, trays, leisure items (toys, flotation devices, marine buoys, canoes, sailboards, sports goods, etc.), industrial parts (business machines, toolboxes, trash containers, hot-water tanks, etc.), and so on. Examples of products are shown in Figures 6.3 to 6.16. Methods of processing these and other products will be reviewed.

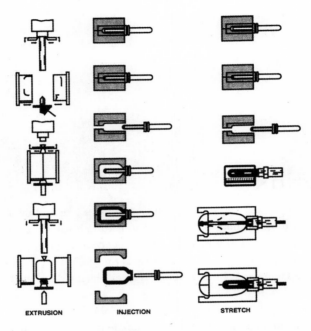

Figure 6.1 Examples of extrusion, injection, and stretch blow-molding techniques

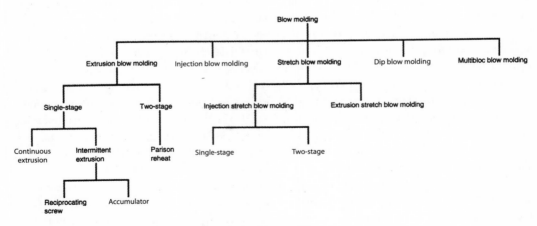

Figure 6.2 Examples of the different forms of blow molding

Although there is more market activity and sales in food and beverage containers, the blow-molded packaging market continues to be active. It consumed 2.5 billion lb of blow-molding resins (25% of the total) in 2001. By 2006, this sector consumed approximately 2.7 billion lb. The largest segments are household and industrial chemical (HIC) containers, drums, intermediate bulk containers (IBCs), tight-head pails (THPs), and motor-oil containers. The HIC market represents 50% of blow-molded industrial packaging. Drums, IBCs, and THPs together amount to 18%. Motor-oil packaging accounts for another 14% of the market. (This study was prepared by Mastio & Co., St. Joseph, Missouri, a well-known consulting firm specializing in industrial-consumer opinion research and market trends in the plastics industry. For more information, call 816-364-6200 or visit www.mastio.com/pt/outlook.html.)

Mastio's report shows that consumption of HDPE far exceeds that of any other resin used for blow-molded industrial packaging. HDPE accounts for nearly 90% of the poundage in this product segment. PET resin, in addition to being used in beverage bottles, has started to appear in bottles for motor oil and other auto and marine fluids. PET is also being utilized more frequently in HIC containers. PET's high clarity imparts greater shelf appeal. Growth for PET in food and beverage containers will average 9% to 10% per year over the next five years, but growth is expected to be only 2% annually for PET industrial blow-molded containers owing to HDPE's lower price and greater chemical resistance. If PET can be price competitive with HDPE, then it is likely that PET will capture market share from HDPE bottles. Polyvinyl chloride (PVC) is losing popularity but is

Figure 6.3 Montage of commercial and industrial blow-molded products

Figure 6.4 Examples of blow-molded foodstuff containers

Figure 6.5 Example of longneck blow-molded products

still used in some industrial packaging. A small volume of automotive and marine fluid bottles and HIC bottles are molded of PVC because it provides hydrocarbon resistance and barrier properties that are not possible with untreated HDPE.

CONTAINER

Large containers ranging from at least 5 gallons (18.9 L) to 55 gallons (208.2 L) are fabricated, continually replacing metal (109). The U.S. Department of Transportation (DOT) continues to

Figure 6.6 Blow-molded containers for potato chips

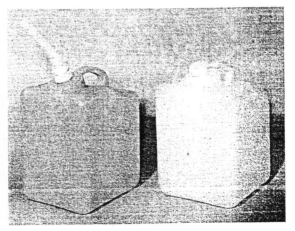

Figure 6.7 Examples of two sizes of blow-molded containers

Figure 6.8 Blow-molded ribbed-panel automotive floor

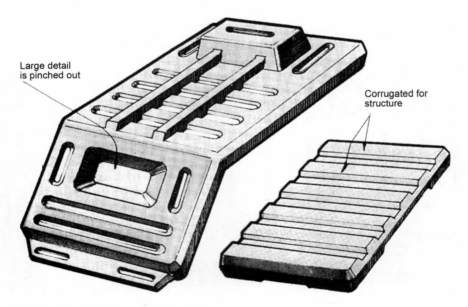

Figure 6.9 Complex 3-D blow-molded products

approve more of these plastic containers for hazardous materials. The light weight of the plastic as well as the number of trips that it can make continues to provide an attractive 55-gallon drum for the industry.

There is no reconditioning necessary, as in steel drums. Plastic containers up to the 55-gallon size can be made on-site as opposed to metal containers that are made and shipped to the consumer. This type of BM is done with accumulator-type BM machines, since the large weight of the parison precludes the use of continuous or intermittent-type machines. The accumulator machines extrude the parison at a high speed so that the parison does not sag.

Blow-molded components have significantly eroded the market for traditional materials, particularly in liquid packaging applications. Since the era of the introduction of polyethylene (PE) squeeze bottles for washing liquids, polyvinyl chloride (PVC) for cooking oils and fruit squash bottles, polyethylene terephthalate (PET) for carbonated beverage bottles, and others, there have been rapid advances in the last few decades in the process machinery as well as the characteristics and range of materials available (chapter 2). Although blow molding of plastics dates from about 1880, rapid developments in machinery and technology only began in the late 1930s, virtually contemporary with the introduction of PE. Blow molding is actually a technique adopted from the

Figure 6.10 Plastic blow-molded fuel tank (left) compared to a metal fuel tank

Figure 6.11 Blow-molded aerodynamic truck wind spoiler

Figure 6.12 Blow-molded 52-gallon hot-water heater that is jacketed by filament winding (chapter 15) to meet UL burst strength requirements

glass industry for the molding of plastic bottles and other containers from thermoplastic materials (chapter 29).

Gasoline Fuel Tank Extrusion blow-molded plastic tanks have been in use by the U.S. military since the 1940s. Gradually, the European and Asian automobiles started using these types of tanks, and later the United States started to adopt them. Different techniques are used to eliminate or reduce gasoline permeability to meet federal standard requirements. They include the use of coextruded structures with barrier plastics, chemically modifying the plastic's surface during or after fabrication (sulfonation or fluorination), and so forth (chapter 20).

Figure 6.13 Blow-molded water flotation wheels

Figure 6.14 Blow-molded swimming pool (courtesy of Vogue Pool Products, La Salle, Quebec, Canada)

Figure 6.15 Blow-molded bellow boots for automotive and other markets

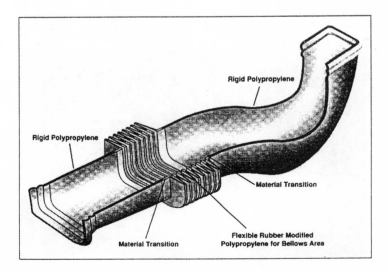

Figure 6.16 Sequential extruded blow-molded polypropylene automotive air duct

In addition to blow molding, there are thermoformed fuel-tank developments. For example, over the past few years, components and systems suppliers, among them Delphi, TI Automotive, Visteon, and Kautex Textron, had interest in thermoforming automotive fuel tanks rather than blow molding them (chapter 7). As reported, recent indications are that thermoforming may have lost some of its luster, while some steel-tank producers are shifting to blow-molded tanks (538). Steel tanks are getting another look, since stricter emissions laws are developing. However, plastic tanks have increasingly gained market share.

Some advantages of thermoforming include its lower capital costs and better wall-thickness control, as well as attachment of secondary components to meet emissions requirements. Advances in blow molding have also occurred. The largest tank manufacturers, Inergy Automotive Systems (Paris) and Kautex Textron (Bonn, Germany), which together control over 65% of the noncaptive tank market, remain firmly in favor of blow molding. Kautex Textron did acquire a twin-sheet thermoforming machine for testing purposes, but the machine has been idle because they are no longer considering thermoforming.

Thermoforming has the limitation of requiring a weld around the complete tank to seal the two halves, resulting in greater emission potentials. They are not as flexible for complex-shaped tanks designed to save space and increase vehicle safety. As TI Automotive reports, with blow molding different designs are being used (vent lines, valves, etc.) so that tanks emissions can be under 20 mg/day compared to the standard blow-molded tanks that emit up to 240 mg/day. TI is also lowering emissions by welding a cover over most of the exterior connections.

An indication that blow-molded tanks will prevail is the reported Fuel Total Systems (FTS) joint venture between Toyota, Horie Metal, and Toyoda Gosei Co. Toyota's vehicles have been using steel tanks almost exclusively. Fewer than 15% of cars made in Japan have plastic fuel tanks. FTS began production during mid-2009 at Horie Metal's plant in Tahara, Japan, and new plant construction started in Lathrop, California.

Industry Size

Blow molding is the third largest plastic processing technique worldwide used for producing many different products. It consumes about 10 wt% of all plastics. Extrusion consumes 36%, injection molding 32 wt%, calendering 8%, coating 5%, compression molding 3%, and others 3%. In the United States alone, there are about 6,000 BM machines, 18,000 extruders, and 80,000 injection-molding machines (IMMs) operating to process the many different types of plastics. U.S. annual sales in BM machines are about $350 million (not accounting for periods of economic downturns).

The BM industry is one of the fastest-growing industries worldwide. The demand for commercial and industrial molding of small to large containers, irregular hollow-shaped industrial products, and bottles is predicted to continue growing. The market for these thermoplastic (TP) products comprises about 22 wt% food, 20% beverages, 15% household chemicals, 12% toiletries and cosmetics, 8% health, 7% industrial chemicals, 5% auto, and 11% others.

BLOW-MOLDING PROCESS

Blow molding can be divided into three major processing categories (Fig. 6.1):

1. Extrusion blow molding (EBM) with continuous or intermittent melt producing a parison from an extruder (chapter 5)
2. Injection blow molding (IBM) with noncontinuous melt from an injection-molding machine (chapter 4) that fabricates a preform supported by a metal core pin
3. Stretched/oriented EBM and IBM to obtain biaxially oriented products providing significantly improved performance-to-cost advantages

Almost 75% of processed plastics are EBM, almost 25% are IBM, and about 1% use other techniques, such as dip blow molding. About 75% of all IBM products are stretched biaxially; there are also stretched EBM products. With stretched blow molding, orientation takes place simultaneously in the hoop and longitudinal directions. These blow-molding processes offer different advantages in producing different types of products based on the plastics to be used, performance requirements, production quantity, and costs (Table 6.1).

BLOWING REQUIREMENTS

Overview The nature of these processes requires the supply of clean compressed air to blow the hot melt located within the BM female cavity. Other gases can be used, such as carbon dioxide (CO_2), to speed up cooling of the blown melt in the mold. Production can increase usually by at least 20%

Extrusion blow molding	*Injection blow molding*
Used for larger parts	Use for smaller parts
Best process for HDPE, PVC and others can be used provided adequate melt strength is available	Best process for GPPS and PP; most resins can be and are used
Much fewer limitations on part proportions, permitting extreme dimensional ratios: long and narrow, flat and wide, double-walled, offset necks, molded-in handles, odd shapes	Scrap-free: no flash to recycle, no pinchoff scars, no postmold trimming
	Injection-molded neck provides more accurate neck-finish dimensions and permits special shapes for complicated safety and tamper-evident closures
Low-cost tooling often made of aluminum; ideal for short-run or long-run production	Accurate and repeatable part weight and thickness control
	Excellent surface finish or texture

Table 6.1 Examples of extrusion vs. injection blow-molding performances

to 40% by using turbulent chilled air at about −30°F that is allowed to escape. This action provides several changes of air through the blow pin during a single blowing cycle.

The gas for blowing usually requires at least a pressure of 30 to 90 psi (0.20 to 0.62 MPa) for EBM and 80 to 145 psi (0.55 to 1 MPa) for IBM. Some of the melts may be exposed as high as 300 psi (2 MPa). Stretch EBM or IBM often requires a pressure up to 580 psi (4 MPa). For IBM of the preform, the pressure is usually 2,000 to 20,000 psi (14 to 138 MPa) and in some cases up to 30,000 psi (207 MPa).

The lower blowing pressures generally create lower internal stresses in the solidified plastics and more proportional stress distribution. The result is improved resistance to all types of stresses (tension, impact, bending, environment, etc.). The higher pressures provide faster molding cycles and ensure excellent conformance to complex shapes.

It is important that dry air be used. Moisture in the blowing air can cause pockmarks on the inside product surface. This defective appearance is particularly objectionable in thin-walled items, such as milk bottles. A system of separators and traps to dry the air taken into the air compressor can prevent this problem.

AIRFLOW CONTROL

Blowing bottles and other products requires air pressures that depend on the plastic being processed (Table 6.2). The air performs three functions: It expands the parison or preform against the closed mold cavity, exerts pressure on the expanding plastic to produce surface details on the cavity, and aids in cooling the parison. Various techniques are used to introduce air into the parison, with the usual going through the extrusion die mandrel (over which the top of the parison has dropped) or the injection mold core pin that supports the preform. In extrusion blow molding, it can also enter from the pinch-off end of the mold or through the side of the mold piercing the parison, usually using a needle that resembles a hypodermic needle (Fig. 6.17). This side action can be located at the parting line of the mold halves.

The type of molding machine available has a strong influence upon the blow opening used. Some machines, for instance, use needle blowing exclusively and excess material is machined off

Plastic	Pressure (psi)
Acetal	100–150
PMMA	50–80
PC	70–150
LDPE	20–60
HDPE	60–100
PP	75–100
PS	40–100
PVC (rigid)	75–100
ABS	50–150

Table 6.2 Examples of air blowing pressure required for certain plastics

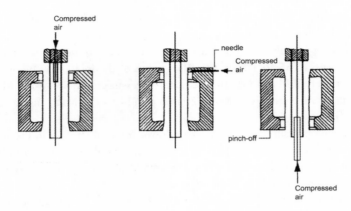

Figure 6.17 Three locations for air to enter extrusion blow molds

to provide larger openings when required. Hollow needle insertion at the mold parting line is considered when a very small opening is required that may even need closing or if the hollow object must be blown up on the side, for example, on carousels with a series of molds. The needle is led right up to the mold cavity by the retainer and moved by a cylinder. This device should be fastened to one mold half. This arrangement should be provided close to a firmly held section of the parison.

During the parison drop between mold halves, the air path is used, allowing for air to be introduced in the parison as it is being extruded to prevent collapse of the parison. This allows for certain functions of blowing products where the parison can be closed after it is cut and inflated for certain processes. The air used for blowing serves to expand the parison tube against the mold walls, forcing the material to assume the shape of the mold cavity and forcing it into the surface details, such as raised lettering, surface designs, and ribs. During the expansion phase of the blowing process, it is desirable to use as high a volume of air as is available so that the expansion of the parison against the mold walls is accomplished in the minimum amount of time.

The aim is to develop a high volumetric airflow at a low linear velocity. A high volumetric flow gives the parison a minimum time to cool before coming in contact with the mold and provides a more uniform rate of expansion. A low linear velocity is desirable to prevent a venturi effect from collapsing a portion of the parison while the remainder is expanding. Volumetric flow is controlled by line pressure and orifice diameter. The flow control valves, which are located as close as possible to the orifice, control linear velocity.

The maximum volumetric flow rate into the cavity at a low linear velocity can be achieved by making the air inlet orifice as large as possible. In the case of blowing inside the neck, this is sometimes difficult. Small air orifices may create a venturi effect, producing a partial vacuum in the tube and causing it to collapse. If the linear velocity of the incoming blow air is too high, the force of this air can actually draw the parison away from the extrusion head end of the mold. This results in an unblown parison. Control valves placed as close as possible to the blow tube must carefully regulate air velocity.

Often, too high a blow pressure will blow out the parison. Too little pressure, on the other hand, will yield end products lacking adequate surface detail. As high a blowing air pressure as possible is desirable to give both minimum blow time (resulting in higher production cycle rates) and finished products that faithfully reproduce the mold surfaces. The optimum blowing pressure is generally found by experimentation on the machinery with the product being produced.

When air contacts the hot plastic, it can result in freeze-off and stresses in the plastic at that point. Air is a material just as is the parison, and as such, it is limited in its ability to blow through an orifice. If the air entrance channel is too small, the required blow time will be excessively long or the pressure exerted on the parison will not be adequate to reproduce the surface details of the mold. A guide to air entrance orifice size when blowing is summarized in Table 6.3.

As reviewed, pressure to be used depends on the plastic being processed. As an example, some PE products with heavy walls can be blown and pressed against the mold walls by air pressures as low as 30 to 40 psi. Low pressure can be used since items with heavy walls cool slowly, giving the plastic more time at a lower viscosity to flow into the indentations of the mold surface. Thin-walled products cool rapidly; therefore, the plastic reaching the mold surface will have a high melt viscosity and higher pressures (in the range of 50 to 100 psi) will be required. Larger products, such as 1-gallon bottles, require increased air pressure (100 to 150 psi). The plastic has to expand farther and takes longer to get to the mold surface. During this time, the melt temperature will drop somewhat, producing a more viscous plastic mass, which in turn requires more air pressure to reproduce the details of the mold.

There is a difference between blowing time and cooling time. Blowing time is much shorter than the time required to cool the thickest section to prevent distortion on ejection. Blow time for an item may be computed from the formula in Table 6.4. Knowing the line pressure and orifice diameter from Table 6.3 will allow you to convert blow time to cubic feet per second. The final mold pressure is assumed to be the line pressure for purposes of this calculation.

During the blowing action, air is heated by the mold, raising its pressure. Calculations ignoring this temperature effect will be satisfactory when blow times are under one second (for small to medium products), but if blow times are longer, the air will have time to pick up heat, resulting in a more rapid pressure buildup and shorter calculated blow time. As shown in Figure 6.18, the blow pin can provide a release channel for the blown air.

Orifice diameter		Product capacity	
(in)	(mm)	(vol.)	
1/16	1.6	Up to one quart	(0.95 liter)
1/4	6.4	1 quart – 1 gallon	(0.95–3.8 liter)
1/2	12.7	1 gallon – 54 gallons	(3.8–205 liter)

Table 6.3 Guide to air entrance orifice size

Gauge (psi)	Orifice Diameter (inch/mm)			
	1/16; 1.6	1/8; 3.2	1/4; 6.4	1/2; 12.7
5	0.993	3.97	15.9	73.5
15	1.68	6.72	26.9	107
30	2.53	10.1	40.4	162
40	3.10	12.4	49.6	198
50	3.66	14.7	58.8	235
80	5.36	21.4	85.6	342
100	6.49	26.8	107.4	429

This is the free air per minute, but since there will be a pressure build up as the parison is inflated, the blow rate of air has to be adjusted.

$$\text{Blow time} = \frac{\text{Mold Vol. Cu. Ft.}}{\text{Ft.}^3/\text{sec.}} \times \frac{\text{Final Mold psi} - 14.7 \text{ psi}}{14.7 \text{ psi}}$$

Table 6.4 Discharge cu ft/s @ 14.7 psi and 70°F with extrusion blow time formula

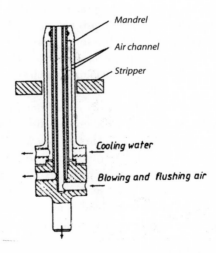

Figure 6.18 Blow-molding pin with escape channel for the blown air

With multicavity molds it is usual to employ a cross-head on which the blowing mandrels are arranged in series. Mandrel chains can also be used for product ejection. The channel for the blowing medium should be as large as possible. Angled radial orifices at the tapered end of the mandrel achieve a fast distribution of the cooling media and turbulence.

The blowing chamber above the neck of the hollow article is known as the lost head. This always becomes a necessity when the width of the neck no longer allows for calibration. In these instances one is concerned with wide-neck containers. A height should be chosen that ensures uniform wall thickness of the neck. A diameter-to-height ratio of 2:1 should be used. When blowing with escaping air, and for fast venting, it is recommended that a tear-off line in the blowing chamber be provided.

EXTRUSION VERSUS INJECTION BLOW MOLDING

Compared to IBM, the advantages of EBM include lower tooling costs, incorporation of blown handle-ware, and so on. Disadvantages or limitations can include the need to control parison swell, the production of scrap, limited wall thickness control, and limited plastic distribution control. If desired, blown solid handles can be molded during the BM process. Trimming can be accomplished in the mold for certain designed molds or secondary trimming operations are included in the production lines. With 3-D molding, scrap is significantly reduced.

With IBM, a main advantage is that no flash or scrap occurs during processing. It gives the best of all wall thicknesses and plastic distribution control, and critical bottleneck finishes are easily molded to higher accuracies. The initial IBM preforming cavities are designed to have the exact dimensions required after blowing the plastic melt as well as accounting for any shrinkage, and so on, that may occur. Neck finishes, internally and externally, can be molded with an accuracy of at least ± 4 mil (± 0.10 mm). It also offers precise weight control in the finished product accurate to at least ± 0.1 g.

Disadvantages can include its high tooling costs and solid-only handle-ware. It was reported in the past that EBMs were restricted or usually limited to very small products; however, large and complex-shaped products were fabricated once the market developed. Similar comparisons exist with biaxial orienting EBM or IBM. With respect to coextrusion, the two methods also have similar advantages and disadvantages, but mainly more advantages for both.

When compared to EBM, the IBM procedure permits the use of plastics that are suitable for EBM, but more importantly, those unsuitable for EBM (other than when certain types are modified). Specifically, it is those with no controllable melt strength, such as the conventional polyethylene terephthalate (PET), that predominantly use the stretch IBM method for large quantities of carbonated beverage bottles (liters and other sizes).

BASICS IN PROCESSING

The blow-molded parts are formed in molds that define the external shapes only. As the name implies, the inner shape is defined by fluid pressure, normally compressed air. In this respect, blow molding differs radically from many molding processes where mold members (male and female cavities are used) determine both inner and outer forms. A major advantage of blow molding is that the inner form is virtually free of constraints because there is no core to extract. The main drawback is that the inner form is only indirectly defined by the mold so high precision and independent internal features, such as in injection molding, are impossible. This has a bearing on wall thickness that can never attain the consistency and accuracy of a full-mold process, such as injection molding (chapter 4).

EXTRUSION BLOW MOLDING

At its most basic, the process involves melt processing a thermoplastic into a tube that is generally referred to as a parison. While still in a heated ductile and firm plastic melt state, the parison is clamped between the halves of a cooled mold so that the open top and bottom ends of the parison are trapped, compressed, and sealed by the mold faces. A blowing tube is also trapped in one parison end, creating a channel through which air pressure is introduced within the sealed parison. Air pressure causes the parison to expand like a balloon so that it takes up the form of the mold cavities (Figs. 6.19 and 6.20). Contact with the cooled mold chills the thermoplastic to its solid state, so the form is retained after the mold is opened and the part removed.

EXTRUDER

The extrusion blow molding process starts by preparing a parison (tube) that is the plastic melt exiting an extruder. Figure 6.21 is a schematic of the plastic melting action in an extruder that has two exiting parisons. This plastic melt has to be a controlled material based on having its basic temperature, pressure, and melting time properly set during its mixing action. Mixing occurs in the extruder's plasticator (screw within a barrel). An example of just setting temperatures in the different zones of a plasticator for different plastics is shown in Table 6.5 As shown in Table 6.6, the plastic output rates of extruders are based on the diameters of screws.

The extruder is usually of the single-screw type. The screw should have a length/diameter (L/D) ratio of 24:1 to 30:1 or more. Screws at the high end are preferred. The compression ratio should be about 3.5:1. This type of information is described with other important information on operating extruders in chapter 5. It is important that one is familiar with the extruder operation in order to ensure that the plastic heated preform is properly prepared to provide blow-molded products that meet performance and cost requirements. Improper preform preparation is damaging to the operation of the blow-molding operation.

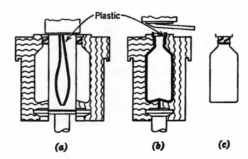

Figure 6.19 Basic processing steps in extrusion blow molding: (a) extruded heated plastic parison, mold open; (b) mold closed and bottle blown; and (c) finished bottle removed from mold

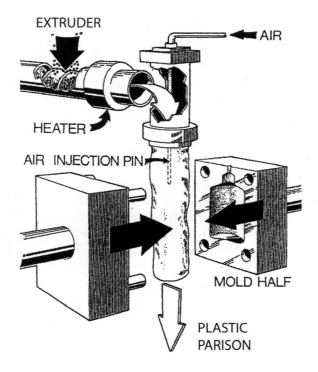

Figure 6.20 Schematic of extrusion blow molding a single parison

MELT FLOW

Understanding and controlling plastic melt flow in the die are important to be successful in blow molding. Factors directly related to die flow include pressure drop, forces acting on the mandrel, and the distribution of melt velocity around the annular flow gap. Die flow also influences the swelling behavior of the melt after it leaves the die. Approximate methods have been proposed to calculate the important aspects of flow in a blow-molding die.

It has been found that the calculation of the major forces and velocity distribution requires only knowledge of the viscous properties of the melt, that is, the viscosity as a function of shear rate and temperature. There have been suggestions that the neglect of the temperature change in the die leads to a large error in the calculated pressure drop in the case of rigid PVC; models proposed for polyolefin flow generally assume the flow to be isothermal (chapter 17).

Such models have been used, for example, to calculate the total force exerted by the melt on the mandrel. For a converging die, the normal pressure acts in a direction opposite to the shear force. This makes it possible to design a die so that the net force on the mandrel is quite small. This simplifies the die design and makes it possible to use a stepper motor rather than a servo hydraulic actuator to drive a movable mandrel for purposes of parison programming.

Use has been made of a power law viscosity equation to calculate the distribution of velocity around the die gap. The resulting model was then used to design dies to provide uniform velocity distributions; dies fabricated based on these designs were found to have very good flow distributions.

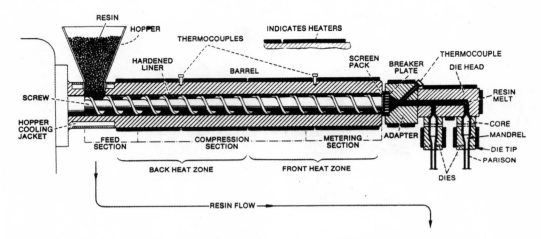

Figure 6.21 Schematic of the plastic melting action in an extruder that has two exiting parisons

Material	Zone 1 temp °F	Zone 2 temp °F	Zone 3 temp °F	Zone 4 temp °F	Melt temp °F	Mold temp °F
ABS	380–440	390–440	400–440	400–440	400–440	110–170
ACRYLIC	370–460	380–460	390–460	400–460	400–460	130–200
LDPE	320–340	320–350	330–360	340–370	340–380	40–100
HDPE	350–390	360–400	370–410	370–420	370–420	40–100
NYLON 6	440–480	450–480	460–480	470–490	450–510	140–200
NYLON 6/6	470–550	480–540	480–540	490–530	510–540	140–200
PC	440–480	450–480	460–480	470–490	460–510	140–220
PC/ABS	420–445	430–455	440–465	450–475	450–490	120–220
PP	340–430	350–440	360–450	370–450	390–450	40–120
PS	340–360	360–390	390–420	400–440	360–440	40–120
PS HIGH IMPACT	370–390	390–420	420–450	430–470	420–460	40–120
PVC FLEXIBLE	250–290	260–300	270–310	280–320	310–375	50–100
PVC RIGID	280–300	300–320	320–350	340–370	340–390	50–100

Suggested starting points only

Table 6.5 Example of temperature conditions in an extruder plasticator based on processing different plastics

Diameter		2½ inch (64 mm)	3½ inch (89 mm)	4½ inch (114 mm)	6 inch (152 mm)	8 inch (203 mm)
LLDPE	Pounds per hour	275–340	550–650	900–1100	1600–1950	2840–3475
	Kg per hour	125–155	250–300	400–500	725–890	1290–1580
LDPE	Pounds per hour	475–575	925–1125	1530–1870	2700–3300	4800–5900
	Kg per hour	210–260	420–510	695–850	1225–1500	2200–2675
HDPE	Pounds per hour	330–425	650–800	1080–1320	1920–2350	3400–4170
	Kg per hour	150–195	300–360	490–600	870–1065	1550–1895
PP	Pounds per hour	360–440	700–865	1170–1430	2080–2540	3700–4500
	Kg per hour	160–200	320–390	530–650	945–1150	1680–2050
FPVC	Pounds per hour	415–510	815–1000	1350–1650	2400–2930	4260–5200
	Kg per hour	190–230	370–450	610–750	1090–1330	1940–2370
RPVC	Pounds per hour	220–270	435–530	720–880	1280–1560	—
	Kg per hour	100–120	195–240	325–400	580–710	
ABS	Pounds per hour	360–440	700–865	1170–1430	2080–2540	3700–4500
	Kg per hour	160–200	320–390	530–650	945–1150	1680–2050
HIPS	Pounds per hour	490–615	965–1200	1600–2000	2840–3550	5050–6320
	Kg per hour	225–280	440–550	725–900	1290–1615	2300–2870
Acrylic	Pounds per hour	415–510	815–1000	1350–1650	2400–2930	4260–5200
	Kg per hour	190–230	370–450	610–750	1090–1330	1940–2370
Poly-carbonate	Pounds per hour	285–350	560–680	925–1125	1650–2000	2900–3500
	Kg per hour	130–160	250–310	420–510	750–900	1325–1600

Table 6.6 Examples of extruder output rates based on processing HDPE

Thus, the forces acting in the die and the velocity distribution are governed by the viscosity function, and variations in temperature need to be taken into account, at least for polyolefins with melt indexes in the range of 0.5 to 1.5 (chapter 3).

Parison Formation Die design influences the behavior of the parison that is formed at the die lips. First, if there is a spider to center the mandrel, weld lines will be formed as the melt flows around the legs (chapter 17). The number of molecules bridging the weld lines to return the melt to a homogeneous state increases rather slowly with time, and for this reason, there will be some lateral weakness in the parison and in the molded product in the neighborhood of a weld line.

Another way in which die flow influences the parison properties is through the process of shear modification or shear refining. The effects of this process are most pronounced in highly branched and high molecular weight linear polyolefins. The effect can be particularly pronounced in the case of blends. The molecular mechanisms underlying the process are not fully understood, but shearing at high rates alters the structure of the melt, decreasing the strength of the interactions between molecules, a process sometimes referred to as disentanglements. The effect is a reversible one, and if a shear modified plastic is annealed for a period of time at a temperature above its melting point, it will recover its preshear structure. However, the time required for this recovery is often much longer than the time a parison is formed and the time it is inflated. Thus, the melt that forms the parison and is inflated into the mold may be a material highly altered by shear in the die. Shear modification reduces elasticity and can result in lower viscosities, especially in the case of branched polymers.

Parison Swell The die flow influences parison behavior through molecular orientation, which manifests itself at the die exit as extrudate swell. Flow at the inlet of a die, where the streamlines are converging rapidly, involves a high rate of stretching in the flow direction. This produces a high degree of molecular orientation, and if the melt is permitted to exit immediately, as for flow through an orifice plate, there would be a very high degree of swell. However, if a long straight section, for example a capillary or straight annular die, follows the entrance, then molecular relaxation processes will lead to a disappearance of the orientation generated at the entrance, and as the die is made longer, the degree of swell is reduced. At the same time, however, the shearing in the die produces some axial orientation, and for a very long die, there is still significant swell resulting from this shear flow. If the die has an expansion or contraction section, this will introduce stretching, and this will introduce orientation whose direction depends on the details of the die design. For these reasons, parison swell is very sensitive to die design.

The melt leaving the die may exhibit sharkskin or melt fracture, which are irregularities in the surface of the extrudate that can affect the surface finish of an extrusion blow-molded container. This effect occurs above a critical shear stress in the die and is often the factor that limits the speed of an extrusion process. Extrudate distortion is most severe in the case of narrow molecular weight distribution (MWD) (chapter 1) high-viscosity plastics. Increasing the temperature or reducing the extrusion rate can sometimes eliminate it, but either of these actions will increase the cycle time. The use of an internal heater in the mandrel is a method to reduce the tendency toward melt fracture on the inside of the parison. The detailed origins of this phenomenon are not fully understood, but the shape and material of construction of the die and the formulation of the plastic are known to be contributing factors.

The cross-sectional shape and surface appearance of an extrudate are governed by many factors in addition to the dimensions of the die lips, and there is currently no reliable method for predicting these characteristics for a given die, resin, or operating condition. A further complication is that as soon as melt leaves the die to form part of a parison, it becomes subject to the force of gravity, which leads to sag or drawdown. This tends to make the parison smaller at the top than at the bottom. The shape of the parison at the moment of inflation is thus the result of the simultaneous processes of sag and swell.

Because of the complexity of this situation, together with the complexity of the rheological properties of the melt (chapter 1), it is not possible to design a blow-molding machine to produce prescribed parison shapes and dimensions at the moment of inflation. However, these dimensions are of crucial importance as they govern the thickness distribution in the finished products. For this reason, machines being used to produce large or irregularly shaped objects are usually equipped with parison programming devices to alter the geometries of the dies (annuli) during parison extrusion to produce parisons that have the prescribed shapes and dimensions. However, parison programming cannot compensate for plastic properties that are basically unsuited for the process, and for this reason, it will be useful to examine in more detail the processes of parison swell and sag.

The wall thickness of the BM product can be related to the swelling ratio of the parison. Referring to Figure 6.22, the swelling thickness of the parison is given by $B_t = h_p/h_d$ and the

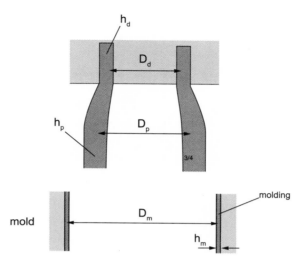

Figure 6.22 Relating thicknesses of swell ratio of parison and BM product

swelling of the parison diameter as $B_p = D_p/D_d$. Using the relationship $B_t = B^2_p$, it follows that $h_p = (h_d)(B^2_p)$. Thus the swell ratio B_p depends on recoverable strain that can be measured (69).

Examples of average parison swell for some plastics are given in Table 6.7.

Both the quality and cost of a blow-molded container are strongly dependent on the parison swell ratios. If the diameter swell is too small, incomplete handles, tabs, or other asymmetrical features may result. On the other hand, if the diameter swell is too large, polymer may be trapped in the mold relief or pleating may occur. Pleating, in turn, can produce webbing in a handle. Swell governs a product's weight and thus the material cost of the molded product. What is desired is the minimum weight that provides the necessary strength and rigidity (Fig. 6.23).

Because swell is a manifestation of the viscoelasticity of the melt, it is time dependent. For example, for high-density polyethylene at 170°C (338°F), 70% to 80% of the swell occurs during the first few seconds after the melt leaves the die, but the remainder can occur over a period of two

Plastics	Swell, percent
HDPE	15–40
LDPE	30–65
PVC (rigid)	30–35
PS	10–20
PC	5–10

Table 6.7 Examples of plastic melt parison swell

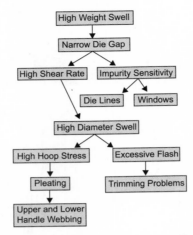

Figure 6.23 Problems encountered in "countering" high-weight swell

to three minutes. For polypropylene at 190°C (374°F), only 50% of the swell occurs in the first few seconds, while more than ten minutes are required to reach an ultimate value. Another source of time dependency in the case of intermittent extrusion is the so-called cuff effect.

The first melt to be extruded after a no-flow period has had an opportunity to relax in the die and experiences relatively little shear just prior to being extruded. Thus, there is less molecular orientation, and thus less swell, than for melt that has just flowed through the entire length of the die. Due to the combined effects of the cuff and the natural viscoelasticity of the melt, the swell is least at the bottom of the parison (the cuff), increasing to a maximum and then decreasing. Of course, sag will accentuate the decrease in swell toward the top of the parison.

Swell increases as the temperature decreases, although it takes place somewhat more slowly. It has been suggested that this effect might be used to control swell by the manipulation of the power for the die heater. Since the time during which the parison is exposed to the air before the mold closes is rather short, and the air surrounding the parison is warm and relatively still, the decrease in temperature that occurs during parison formation is generally rather small, usually less than 5°C (9°F). For this reason, parison formation is generally assumed to be an isothermal process.

Swell increases as the flow rate increases, owing to the enhanced orientation resulting from molecular orientation in the die. Swell varies greatly from one plastic to another and is strongly affected by the extent of branching and the molecular weight distribution. For linear plastics, a broader MWD generally results in a higher swell. However, plastics with very similar measured molecular weight distributions can have significantly different swell behavior, and this probably reflects the fact that swell is highly sensitive to amounts of high molecular weight material that are too small to be detected using standard analytical methods. Highly branched plastics tend to swell more, but it is not possible to generalize when both branching and MWD are changing.

Because swell is an elastic recoil process that results from molecular orientation in the die, the shape of the die channel has a strong effect on both diameter and thickness swells. The simplest case is a long straight annular die. Here, we have only shear flow with no stretching and we would expect to see some orientation in the axial direction. This suggests that there would be no preferential direction of swell and that the diameter swell should be equal to the thickness swell.

For the more complex BM dies, the swell ratios are strongly influenced by two die geometry features, the angle of divergence or convergence of the outer die wall and the variation of gap spacing along the flow path. A diverging die stretches the melt in the hoop direction, and this should reduce diameter swell by counteracting the axial orientation generated by the shear flow.

Basics of Melt Flow The non-Newtonian behavior of a plastic (chapters 1 and 17) makes its flow through a die somewhat complicated. When a plastic melt is extruded from the die, there is the usual condition of swelling (Fig. 6.24 and Table 6.8). After exiting the die, it is usually stretched down to a size equal to or smaller than the die opening. The dimensions are reduced proportionally so that, in an ideal plastic, the drawn-down section is the same as the original section but smaller proportionally in each dimension. The effects of melt elasticity mean that the material does not draw down in a simple proportional manner; thus the drawdown process is a source of errors in the profile. The errors are significantly reduced in a circular extrudate, such as in a BM parison. These errors must be corrected by modifying the die and take-off equipment (5).

There are substantial influences on the material because of the flow orientation of the molecules, so there are different properties parallel and perpendicular to the flow direction. These differences have significant effects on the performance of the products.

Another important characteristic is that melts are affected by the orifice shapes. The effect of the orifice is related to the melt condition and the die design (land length, etc.). A slow cooling rate can have a significant influence, especially with thick parts. Cooling is more rapid at the corners; in fact, a hot center section can cause a part to "blow" outward and/or include visible or invisible vacuum bubbles.

Parison Sag

Sag (drawdown) can cause a large variation in thickness and diameter along the parison, and in an extreme case, can cause the parison to break off. For a Newtonian fluid, sag can be kept under

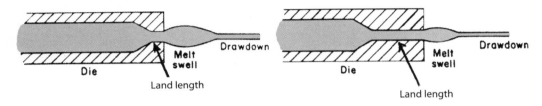

Figure 6.24 Effect of land length on swell

SHEAR RATE, S^{-1}	DIE SWELL RATIO AT 392 F (200 C)			
PLASTIC	10	100	400	700
PMMA-HI	1.17	1.27	1.35	—
LDPE	1.45	1.58	1.71	1.90
HDPE	1.49	1.92	2.15	—
PP—copolymer	1.52	1.84	2.1	—
PP—homopolymer	1.61	1.9	2.05	—
HIPS	1.22	1.4	—	—
HIPVC	1.35	1.5	1.52	1.53

Table 6.8 General effect of shear rate on die swell of various thermoplastics

control simply by using a material with a sufficiently high viscosity. Plastics sag because they are non-Newtonian and generally have melt indexes of less than 1 (chapter 1). Sag becomes more severe as the temperature is increased. Since plastic melts are viscoelastic, resistance to sag cannot be quantitatively correlated with viscosity.

A number of proposals have been made as to which viscoelastic property (chapter 1) of a melt governs sag. There is the linear creep compliance, as calculated from a tensile relaxation modulus, to predict sag. Another suggests that the extensional stress growth function can be correlated with sag behavior. J. M. Dealy (McGill University) points out that the sag process is neither a constant stress nor a constant strain rate process, so neither of these functions is directly relevant (540). Noting that the strain rate is small, it is proposed that the use of a relatively simple general model of a viscoelastic liquid be used.

Reliable methods for predicting sag behavior on the basis of well-defined rheological properties are still lacking. In their absence, empirical techniques for evaluating sagging tendencies are employed. An example is to make video records of parison length versus time. A simpler procedure that might be useful is to determine the extensibility of film plastic by simply clamping a weight to a specified length of extrudate from a melt indexer and monitoring the resulting behavior.

When we consider the combined effects of swell and sag, the situation becomes quite complex from a rheological point of view. Figure 6.25 is a sketch of parison length versus time curves for various cases. The first part of the curve reflects the extrusion period during which the parison is formed. Once extrusion stops, the length is governed entirely by swell and sag. Curve no. 1 shows the case where there is swell but not sag, while curve no. 2 is for the case of sag with no swell. In the case of an actual parison, curve no. 3, there is initial recoil followed by a slower increase in length, reflecting the complex time dependency of the swell.

A number of models of the parison formation process have been formulated. All of these models assume that swell and sag are in some way additive. Unfortunately, none of them can be used to predict with confidence the behavior of the parison for a given die, resin, or operating condition.

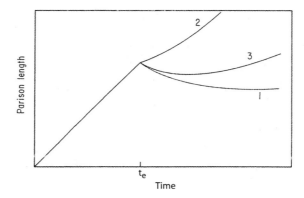

Figure 6.25 Parison length vs. time curves for three different situations

Pleating The action of pleating (also called draping or curtaining) is a buckling of the parison that occurs when the melt at the top of the parison is unable to withstand the hoop stresses due to the weight of the parison suspended from it. Pleating is undesirable because it can cause webbing in the blown product.

Obviously, large diameter swell and a small die gap are factors that will increase the probability of pleating. It is not clear exactly which rheological property governs the ability of a melt to resist pleating, but the viscosity is a rough guide to the melt stiffness that is desired. Thus, high-viscosity plastics are less likely to pleat, and increasing the temperature increases the likelihood of pleating.

Parison Inflation The behavior of the closed parison during the inflation process is a manifestation of its extensional-flow rheological properties. It has been observed that the parison does not inflate uniformly and tends to bulge out in the center. A blowout can occur if the ratio of the mold diameter to the parison diameter (the blowup ratio) is too high.

The deformation is not the same as that which occurs in the uniaxial extension experiment, but the results of such an experiment are still thought to be relevant to inflation performance (18). In particular, it is thought that extension thickening behavior implies that a plastic will be easy to inflate and unlikely to exhibit blowout, even when the blowup ratio and inflation pressure are high. Extension thinning, on the other hand, is thought to imply unstable inflation and an increased likelihood of blowouts. This hypothesis is consistent with the observation that low-density (branched) polyethylene is easier to inflate than HDPE.

Melt Fracture Melt fracture is a loosely defined term that has been applied to various forms of extrudate roughness or distortion encountered at high extrusion rates for all plastic melts. In some cases, there may be small-scale roughness, rippling, or sharkskin, in others a very regular helical screw-thread extrudate. Generally, at very high rates the extrudate becomes completely irregular.

The term *melt fracture* originated from early observations of marker threads or of flow birefringence at the entrance of a die. A line that was continuous at low extrusion rates could be seen to snap suddenly and to recoil when the flow rate was increased, and visible evidence of roughness was noted simultaneously. These observations suggested a mechanism of literal breakage of melt that had been stretched or oriented beyond its natural limit in flow. Other mechanisms that have been proposed include slippage at the die entry or along the die wall and instability of the circulating vortex flow, often seen at the corners of a flat entry inlet to a die. The small-scale roughness has been attributed to stick-slip flow at the die exit.

It should be noted that in some cases the flow curve (viscosity vs. shear rate) is not affected at all by the onset of melt fracture. Some plastics, however, exhibit pronounced discontinuities of the flow curves. The most important of these is linear polyethylene. The onset of visible roughness occurs near, but not necessarily at, the flow discontinuity. The discontinuity has been ascribed to slippage at the wall, leading to increased flow and lower apparent viscosity.

The effect of slippage on the extrudate depends on the nature of the extrusion apparatus. A very stiff apparatus, such as a piston driven by a motor, will give a constant extrusion rate but severe pressure fluctuations. A soft apparatus, for example a gas-driven rheometer, will produce large variations in output. The most serious effect occurs when the extrusion is done at a rate very close to the critical one for the onset of slippage. The output can then oscillate between the two branches of the flow curve as shown in Figure 6.26, causing large variations in flow rate and gross distortions of the extrudate. The severity of distortion is also markedly dependent on die geometry.

These effects may be rationalized by considering that in the absence of slippage the melt is highly oriented by the flow. When the melt slips suddenly, it creates room for the relatively unoriented stagnant melt that has been circulating in the entry vortexes to enter the die and to be extruded without the usual stretching at the die entry. The alternating of highly oriented extrudate (with large elastic recovery) with the unoriented melt accounts for the extrudate distortion.

Also, there is a pronounced effect of inlet taper angle on the extent of the distortion. A correctly tapered inlet suppresses the formation of a stagnant vortex. Therefore, even if slippage occurs, there is no stagnant material to cause nonuniformity and the external appearance of the extrudate is relatively smooth. It has also been reported that the use of a slightly tapered conical die, rather than a cylindrical one, delays the onset and minimizes the severity of melt fracture.

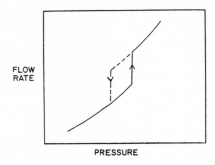

Figure 6.26 Oscillating melt flow rate near slip discontinuity of flow curve

It is generally agreed that shear stress is a better criterion for the onset of melt fracture than shear rate. For example, when the extrusion temperature of an LDPE is varied from 130° to 230°C (266° to 446°F), the shear stress at the onset of melt fracture varies by only about 30%, whereas the critical shear rate varies by a factor of 100. The critical stress generally ranges from about 10^6 dynes/cm^2 (14.5 psi) for a nonpolar polymer (such as polystyrene, polypropylene, and polyethylene) to about 10^7 dynes/cm^2 (145 psi) for a polar polymer, such as nylon, polyacetal, and PET.

The effect of molecular weight and MWD is not as clear-cut. The most comprehensive study indicates that the critical stress increases linearly with the reciprocal of the average molecular weight; that is, the critical stress is lowest for high molecular weight polymers.

The effect of MWD on the critical stress is less well defined, with many conflicting reports in the literature. Because of the difficulty of measuring the extremes of the MWD, it is probably not practical to use such a correlation for quantitative predictions in any case. As with die swell, it is easier and more reliable to measure the critical shear stress for the polymer of interest.

Extensional Flow Historically, shear flow is important because it is the simplest way to deform a fluid plastic, particularly a low-viscosity fluid plastic, in a controlled manner that can be analyzed and used to measure the viscosity. Of course, it is a flow that occurs in nearly all plastic melt processing operations. However, shear flow is not the only deformation that can be applied to a fluid. In general, the flow that occurs in melt processing machines may be quite complex.

In addition to simple shear flow, extensional flow usually occurs. The most elementary type of extensional flow is viscosity, and end correction is needed in any case to characterize the processing parameters, thus specifying the die swell for any extrusion condition. It should be noted, of course, that a die swell curve depends upon the molecular weight and distribution of the plastic, as do rheological properties in general.

Next is a review on the dependence of die swell on molecular parameters where the situation is not completely clear. With one exception, the literature indicates that die swell increases with increasing breadth of MWD. The breadth of the MWD is usually specified in terms of a ratio of two types of average molecular weight. However, these averages are strongly affected by the shape of the MWD, especially at the extremes of high and low molecular weight. It has been found that a blend containing a very low molecular weight fraction shows unexpectedly high swell. Conversely, it has been found that the addition of as little as 0.1% of an ultrahigh molecular weight polyethylene increases the die swell of an HDPE.

Perhaps it is best to say that die swell varies with the MWD in a manner not predictable by a simple parameter but depends in some complicated way on the entire shape of the MWD. Another possible complication is that an HDPE may contain small amounts of long chain branching (LCB), and the amount can vary with the details of the polymerization process. It is known that LCB has a strong influence on melt rheology, especially elasticity.

Fillers, particulate or fibrous, have the effect of reducing or suppressing die swell. Small amounts of crystallinity have similar effects, and this may account for the surprisingly small die swell of rigid polyvinyl chloride.

Parison Head

The parison head, sometimes called the die head or simply the die, is a specialized form of tube extrusion die (chapter 17). Its function is to deliver a straight parison in the correct diameter, length, and wall thickness and at the correct temperature for blow molding. Prior to being clamped in the mold, the parison is suspended unsupported in free air. To avoid undue deformation, it is necessary to extrude the parison vertically, downward.

The blow-molder extruder is almost always arranged in a horizontal attitude, so the first task for the parison head is to turn the melt flow stream through a right angle. This is basically difficult and undesirable but necessary to achieve in a way that meets the essential requirement for a constant flow rate at every point in the annular die gap. A second and related requirement is that the parison should carry as little evidence as possible of the weld line (chapter 17) formed when the melt stream from the extruder flows around the torpedo. Many parison die-head designs have been evolved to deal with these problems. An example, known as the heart-shaped cross-head, is shown in Figures 6.27 and 6.28. Figure 6.27 provides a simplified view. More details are given in Figure 6.28; this figure's call-outs are as follows: 1 = housing, 2 = torpedo, 3 = seating, 4 = melt flow channel, 5 = die, 6 = mandrel, 7 = die gap, 8 = retainer ring, 9 and 10 = die gap adjustment means, AG = plane of annular flow channel, and E = flow divider edge. Head design has benefited

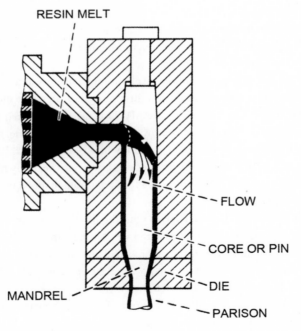

Figure 6.27 Simplified view of a heart-shaped parison die head

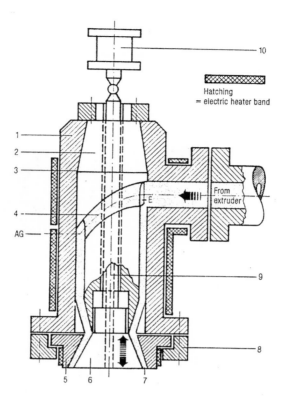

Figure 6.28 Details of a heart-shaped parison die head

from computer analysis of the internal flow channels, which has removed much of the trial and error and given a greater assurance of success.

In addition to Figure 6.27 and Figure 6.28, Figures 6.29 to 6.32 show different continuous parison die heads that provide different capabilities to minimize or eliminate weld lines.

PARISON WALL THICKNESS

A potential blow-molding problem is one shared with pipe and tube extrusion dies (chapter 17). This is the difficulty of ensuring that the mandrel (also called core or torpedo) remains coaxial with the die, without dividing the melt stream with mechanical supports or too much excessive wall thickness deviation. The parison die heads are examples of different design approaches. Those without supports (no spiders) in the melt stream rely instead on accurate seating between the mandrel and die. The parts need to be relatively massive for this design to be sufficiently rigid. Smaller torpedoes will need supports in the melt stream and here the challenge is to minimize the formation of weld lines in the parison. Such supports include two spider legs, staggered spider legs, breaker plates, screen tubes, and most usefully, spiral mandrels (chapter 17). The problem is a difficult one and in the worst case, the weld lines show up as local variations and/or weaknesses in the parison wall thickness.

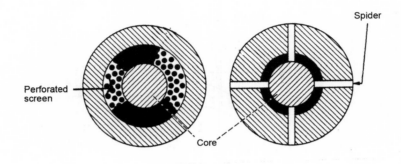

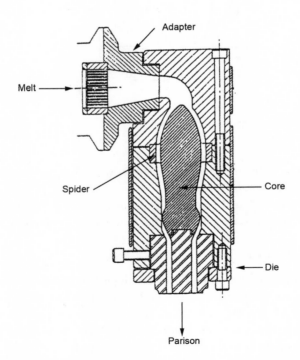

Figure 6.29 Side view of center-fed die with spider supports for its core; top view: examples of four-spider support system or use of a perforated screen

The basic parison is a tube that in principle at least has a relatively constant wall thickness at all points around its circumference and along its length. In practice, circumferential thickness may vary as a result of weld lines or if the mandrel is not truly coaxial with the die. Thickness along the parison length may vary because of tensile thinning caused by the weight of the dependent parison and/or uneven melt flow from the extruder as well as restrictions in the parison die head. However, the ideal parison with a constant wall thickness is not necessarily the optimum for blow molding. Even simple molded shapes encompass considerable variations in profile.

For example, the body of a bottle is much greater in diameter than the neck. For some containers, there may be several significant variations along the axis, and the effect is usually pronounced in technical blow moldings. If such articles are blow molded from a parison of constant wall thickness, then the finished molding will be thinner where the parison has expanded the most and thicker where it has expanded the least. In most cases, the ideal is a finished product with a constant wall thickness at all points. To approach this ideal, what is used is a parison in which the wall thickness varies along the length so that it is thickest where it must expand the most. This is achieved by parison programming or profiling.

As shown in Figure 6.33, the usual way to do this is by moving the mandrel in an axial direction relative to the die. If both die and mandrel are provided with conical outlet features, this movement will increase or decrease the annular die gap between the two. A servo system acting on a preprogrammed thickness profile controls the mandrel movements. The system may include a feedback loop to adjust parison profiling in response to screw speed variations in the extruder.

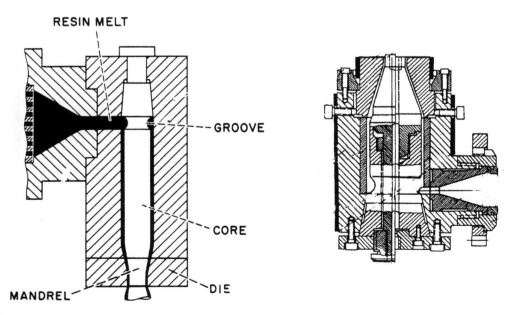

Figure 6.30 Examples of a grooved-core parison die head

- Overlapping layers for optimum roundness and radial distribution
- Easy cleaning
- Constant parison flow with multiple parisons (1 ... 10fold)
- Pin adjustment with low adjustment force and low guidance wear
- Only one heating zone, with low self-heating in the distribution channels because of optimized calculation methods
- Rapid head change because of separation between head and parison programming actuation

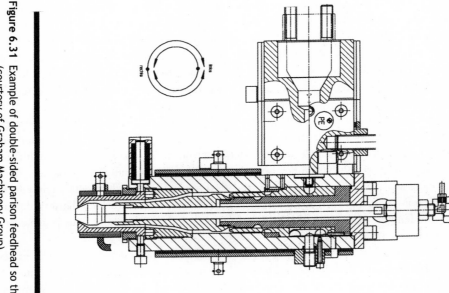

Figure 6.31 Example of double-sided parison feedhead so that a double-layered parison is produced that overlaps weld lines 180° apart (courtesy of Graham Machinery Group)

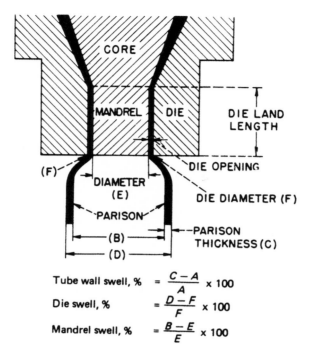

Figure 6.32 Explanations of a parison die head

There are two types of mandrel and die bushings. One is the convergent type and the other is the divergent type (Fig. 6.34). The selection of which mandrel and die bushing (commonly known in blow molding as head tooling) is made by the size of the parison necessary. Convergent tooling is the easiest to control and is used whenever possible. Divergent tooling is used when the parison needs to be larger so the tooling causes the parison to flare out as it is extruded. Both convergent and divergent mandrels are mounted directly to the programming mandrel and are programmed up and down to open or close the gap between the pin and bushing (Fig. 6.35).

For product shapes that are not circular, variations in the preform thickness perpendicular to the preform at the die opening are designed to provide the final dimensions. Figure 6.36 is an example of a rectangular shape.

Machine Design

In extrusion blow molding, an extruder feeding a parison head is used to produce a parison as a precursor or preform for the molding process. The configuration of individual machines may vary greatly, but some essential elements can be distinguished, such as those shown in Figure 6.37. The

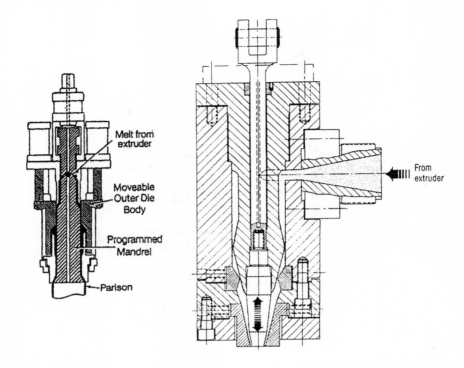

Figure 6.33 Examples of parison wall thickness control by axial movement of the mandrel

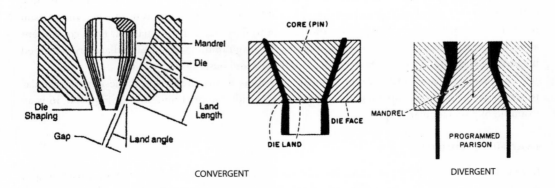

Figure 6.34 Examples of convergent and divergent die-head tooling

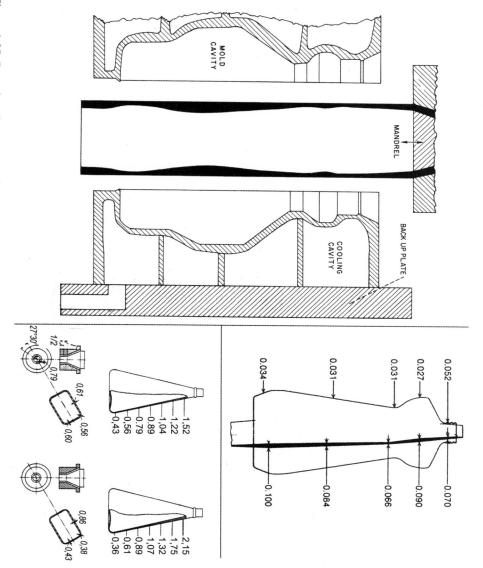

Figure 6.35 Examples of programmed parisons

extruder and parison head are arranged to extrude a parison vertically between the two halves of a blow mold.

The mold halves are clamped to platens that are linked to a mold closing and clamping device. A blow pin is provided to inject air under pressure into the parison. Because blow molding is conducted at relatively low pressures, the construction of the machine and mold can be much lighter than is required for injection molding. Consequently, machines, and particularly molds, cost less than those used in injection molding. Using multiple molds or multicavity molds with multiple

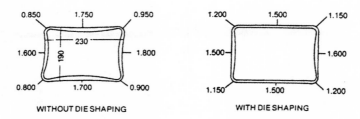

Figure 6.36 Example of rectangular parison shapes where (a) die opening had a uniform thickness resulting in weak corners and (b) die opening was designed to meet the thickness requirements required

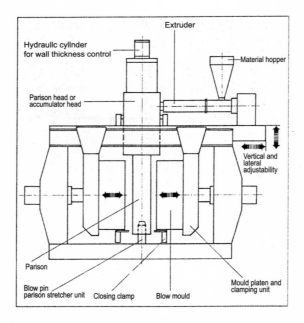

Figure 6.37 Simplified schematic showing parts of a blow-molding machine

parisons can increase machine production rates. These general principles apply to a number of distinct extrusion blow-molding machine designs.

SINGLE-STAGE DESIGN

Single-stage extrusion blow molding produces a blown product in a single integrated process cycle. Blow molding immediately follows parison extrusion and relies on the melt condition of the parison for the deformation and flow necessary to take up the shape of the mold. There is no reheating of the parison before molding. The principal variants are the continuous extrusion and the intermittent extrusion processes. Extrusion in this distinguishing sense applies to extrusion of the parison rather than the operation of the extruder.

TWO-STAGE DESIGN

Two-stage extrusion blow molding treats the parison as a true preform by separating the functions of parison preparation and blow molding. Parisons are produced by conventional tube-extrusion methods (chapter 5). The tube is cooled and cut into parison lengths that are stored before being reheated and blow molded in the normal way (Fig. 6.38).

The reheating process can be selective, leading to a degree of control over wall thickness in the blown product. The reheating process also tends to relax stress in the parison, resulting in a stronger product. However, the process has a number of disadvantages. The plastic heat history is increased by the use of two heating cycles, energy use is increased, and costs arise from the storage and handling of parisons. The process is not suited to high production speeds and is little used.

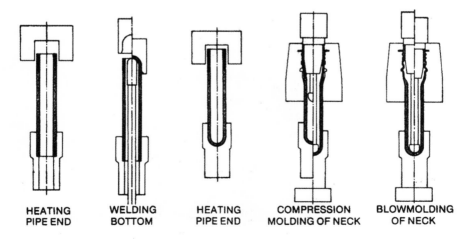

Figure 6.38 Examples of preparing cut-to-size parisons for a two-stage extrusion blow-molding process (courtesy of SIG Plastics International)

CONTINUOUS EXTRUSION DESIGN

In this process, the parison is continuously extruded between the open mold halves from an accumulator head. When the required length of parison has been produced, the mold is closed, trapping the parison that is severed by a hot knife from the die. Figure 6.39 provides a simplified schematic of a continuous blow-molding process. More details are given in Figure 6.40.

The continuous EBM (Fig. 6.40) shows two different techniques for blow molding a container with a calibrated neck. The top schematics (a and b) illustrate a top mold closing system where a = mold, b = calibrating blow mandrel, c = parison, d = parison head, e = extruder, f = neck insert, g = pinch-off blade, h = pinch-off, and i = clearance for scrap waste. The bottom schematic (a, b, and c) illustrates a movement by the mandrel where a = blow mold, b = calibrating blow mandrel, c = parison, d = parison head, e = extruder, f = cutting insert in mold, g = scrap waste, and h = pinch-off.

Land or pinch-off areas on the mold compress and seal the upper and lower ends of the parison to make an elastic airtight part. Compressed air is introduced through the blow pin into the interior of the sealed parison that expands to take up the shape of the mold cavities. The cooled mold chills the blown object that can then be ejected when the mold opens.

The blowing air can be introduced in a variety of ways. In the simplest case, the parison is extruded downward so that the open end slips over a blow pin. Alternatively, the blow pin can be introduced from above the mold after the parison is severed, or blow needles can be built into the mold, where they pierce the parison as the mold closes.

Shuttle Relative movement between the parison head and mold is necessary so that parison extrusion can proceed continuously while the mold is closed. This is achieved in many different ways. The mold may be lowered, moved aside laterally, or swung aside on an accurate path. Alternatively, the extruder may be moved while the mold remains stationary. Two or more molds can be used

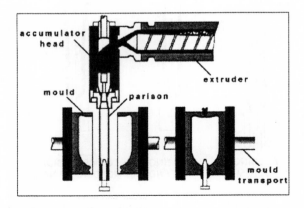

Figure 6.39 Introduction to a continuous extruded blow-molding system with its accumulator die head

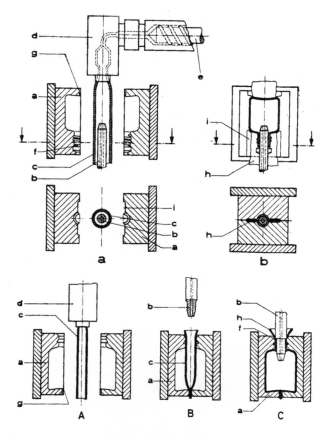

Figure 6.40 Examples of continuous extruded blow-molding systems with calibrated necks

in a shuttle arrangement, so that one or one set is open for parison extrusion while the others are performing the blowing and cooling cycle (Fig. 6.41 to Fig. 6.47; 539).

Rotary There is a method of mold rotary movement that results in very high production rates. In such machines, known as horizontal or vertical wheel machines (also called carousel or ferris wheel machines), a number of molds are mounted on the machine's rotary table. Movement of the table carries the closed mold away and presents a new open mold to the die head, allowing extrusion to continue (Fig. 6.48 to Fig. 6.52).

Except in special cases such as the shuttle and wheel machines, the extrusion rate of the parison must be synchronized with the blow cycle for a single mold. This may involve a relatively slow extrusion rate, running the risk of thinning as the parison stretches under its own weight. For this reason, continuous extrusion blow molding is best suited to thermoplastics with a high melt viscosity or high melt strength, or to short or thin-walled parisons. Certain plastics, such as polypropylenes, have container capacities at upper limits of of about 10 liters.

Intermittent Extrusion Design

In intermittent extrusion blow molding, the parison is extruded from the parison die head discontinuously. When a parison of the required length has been produced, extrusion is interrupted

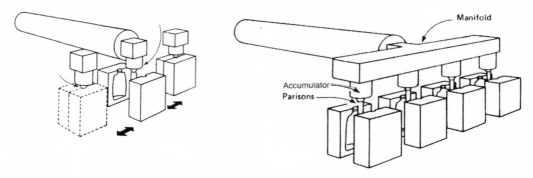

Figure 6.41 Schematics of continuous two-mold and multimold shuttle systems

Figure 6.42 View of a three-milk bottle mold shuttle system

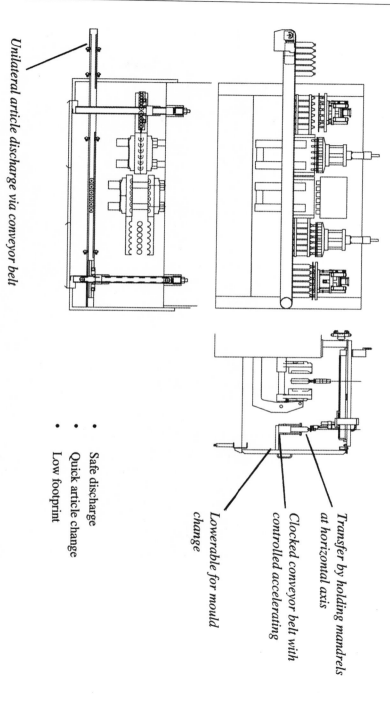

Figure 6.43 Schematic of dual-sided shuttle with six parisons (courtesy of Graham Machinery Group)

- Safe discharge
- Quick article change
- Low footprint

Figure 6.44 Closeup of dual-sided shuttle with six parisons (courtesy of Graham Machinery Group)

Figure 6.45 Dual-sided shuttle with six parisons with safety doors opened (courtesy of Graham Machinery Group)

Figure 6.46 Dual-sided shuttle with six parisons with safety doors closed (courtesy of Graham Machinery Group)

Overcoming Shuttle Machine Limitations

- Another possible approach is to utilize not two, but four molds, somehow shuttling the molds along multiple axes.
- This is also not reasonable, due to the impracticality of adding such mechanical complexity to the machine.

The Rotary Shuttle Machine

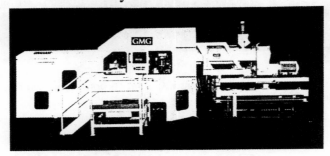

A Concept in Blow Molding

The Rotary Shuttle Concept

- Incorporate four molds by utilizing a rotary axis of movement
- Use a single flowhead for four molds
- Shuttle the molds perpendicular to the axis of rotation to avoid hitting the flowhead
- Index in 90° increments

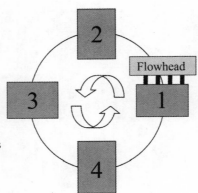

Figure 6.47 Overcoming shuttle machine limitations (courtesy of Graham Machinery Group)

Station Number 1

- This is the station where the molds capture the parison

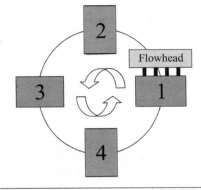

Station Number 1

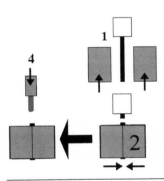

1. As the wheel indexes 90 degrees, the open mold moves to a position under the flowhead
2. Molds close to capture parison
3. Knife cuts parison and molds shuttle back before next indexing step
4. Blow pins enter molds after mold shuttles a short distance to the home position
5. Molds index 90 degrees

Stations 2-3

Extended Cooling

- Cooling begins as soon as the blow pins enter at station 1
- Cooling ends as the molds open in station # 4

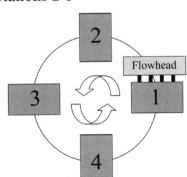

Figure 6.47 *(continued)*

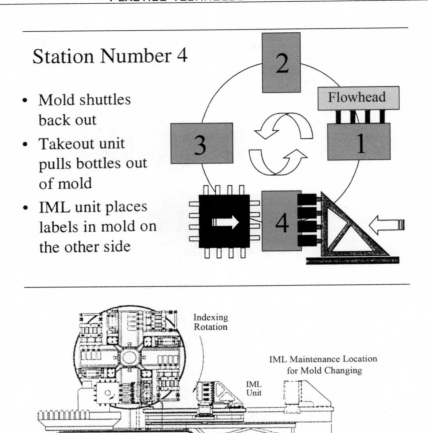

Figure 6.47 *(continued)*

until the next parison is required for the subsequent blowing cycle. This makes it unnecessary to have relative movement between the die and mold. The process is generally used to make larger blow moldings, such as drums or automotive fuel tanks.

These products require large and heavy parisons, which when extruded normally, sag and thin down under their own weight. Intermittent rapid extrusion reduces this tendency, and it is achieved either by reciprocating screw or accumulator means. Intermittent extrusion blow molding is best suited to long or heavy parisons and to thermoplastics with a low melt viscosity or low melt strength.

Reciprocating Screw Reciprocating screw intermittent extrusion blow-molding machines use extruder units in which the screws are capable of axial as well as rotational movement. The function is identical to that of the injection unit on an injection-molding machine (Fig. 6.53). The injection-molding machines described in chapter 4 are used. This intermittent extrusion blow-molding machine operates as an injection-molding machine except at a lower melt pressure and flow rate.

The parison is prepared in two steps. The first step is melt preparation. As explained in chapter 4, the screw rotates, thus heating and melting the material. It is conveyed along the screw flights to

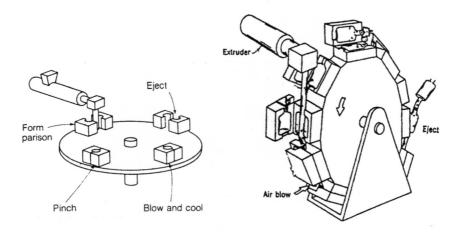

Figure 6.48 Schematics of continuous horizontal or vertical wheel machines

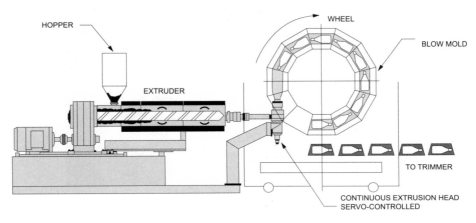

Figure 6.49 Schematics of vertical wheel machine in a production line (courtesy of Graham Machinery Group)

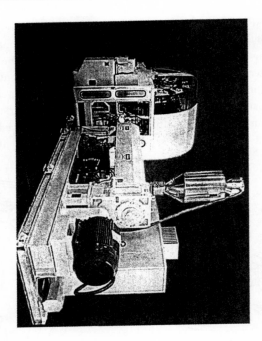
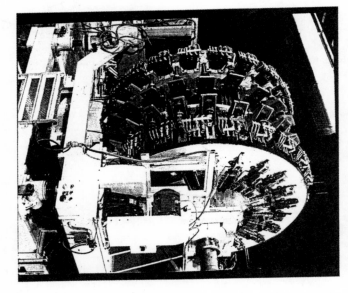

Figure 6.50 Rotary machine with closeup of rotary wheel (courtesy of Graham Machinery Group)

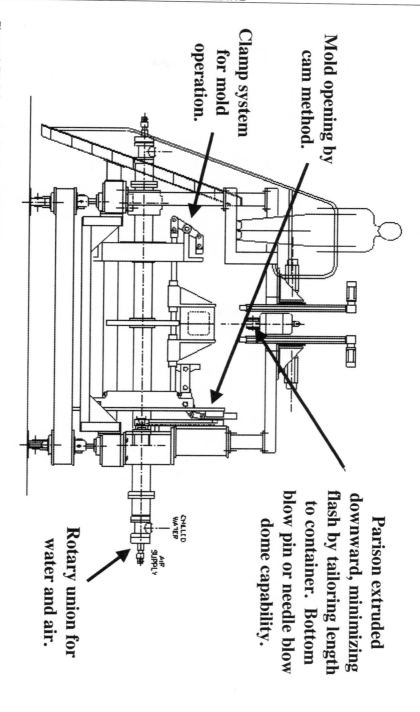

Figure 6.51 Schematic side view of five-station rotary wheel (courtesy of Graham Machinery Group)

Rotary Shuttle Advantages

- We will compare performance to a linear shuttle machine
- Each machine to have 16 cavities, producing bottles on 100 mm center distances

Shuttle Machinery Production

- "Typical" machines: 1-6 parisons
- "Long-Stroke" machines: 8, 10, 12 or more parisons.
- The name "long-stroke" refers to the large axial distance the molds must "shuttle" through to move from the flow heads to the blow pin station.

Flowhead Comparison

Long Stroke Shuttle
- Requires 8 parisons.
- As more parisons are added control adjustments become more complicated. This is especially true in multilayer applications.
- Distribution blocks and flowheads are expensive.

Rotary Shuttle
- Provides 16 cavity output with only 4 parisons.
- Improves bottle weight consistency.
- Reduces flow head complexity and cost.

Figure 6.52 Rotary shuttle advantages (courtesy of Graham Machinery Group)

Long Stroke Shuttle

Rotary Shuttle

Mold Comparison

Long Stroke Shuttle
- Two large eight-cavity molds

Rotary Shuttle
- Four smaller four-cavity molds

> The rotary shuttle is versatile- it can accept existing shuttle molds. This allows converters to run small runs using two molds on a small shuttle machine, then ramping up to four molds on a single rotary shuttle machine if orders increase.

Figure 6.52 *(continued)*

Cooling Efficiency

- The shorter shuttling distance of the rotary shuttle allow a larger percentage of the cycle to be used for cooling.
- For example, if a cooling time of 11 seconds is required, the overall cycle might be about 16.25 seconds on a linear shuttle, and 15 seconds on the rotary shuttle.

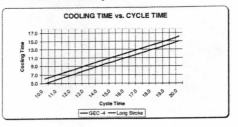

In Mold Labeling – Long Stroke Shuttle

- Long horizontal motion- the IML heads must travel the entire length of the mold horizontally, slowing cycle time significantly. Total cycle time delay of 3-5 seconds typical.
- Label placement accuracy may be compromised due to vibrations in the cantilevered IML unit.

In Mold Labeling – Rotary Shuttle

- Short horizontal motion- the IML heads move into position as the molds index, and travel a short distance. Cycle time delay of less than 0.5 seconds per index, or 2 seconds for complete cycle time.
- The IML heads can actually move into the mold before the bottles are fully extracted.
- High label placement accuracy.

Figure 6.52 *(continued)*

In-Mold Labeling Unit

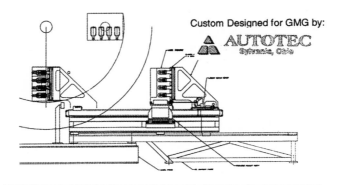

Footprint: Rotary Shuttle vs. Long-Stroke

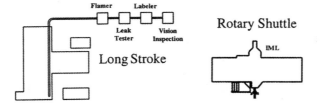

Comparison of equipment and space required to produce labeled, leak tested bottles including vision inspection

Comparison of Tooling Components Required for 16 Cavity Moldset

Tooling Component	"Long stroke"	GEC-4
Die pin and bushing Sets	8	4
Cut-off knives	8	4
Blow pins	16	16
Transfer holders / assemblies	2	1
Trim dies and punches	16	4
Leak tester heads	If available - 16	4
IML Machine Units	2	1
IML label baskets	32	8
IML deployment heads	32	8

Figure 6.52 *(continued)*

Summary: Rotary Shuttle vs. Long-Stroke Shuttle

- Fewest parisons for cavitation, less tooling required
- Existing shuttle molds can be utilized
- Very low cycle time penalty for IML
- Fully integrated with:
  - Take-out
  - Deflash trimmer
  - Detabber
  - Leak tester (optional)
  - Vision inspection (optional)
- Compact design requires minimum floor space
- Lowest energy consumption

Current Development

- Clamp force increasing from 9.5 tons to 15 tons
- Platen size increasing
- Integrated "electronic cam" control for higher speed and reduced cycle time capability from 12 seconds to 10 seconds
- Mold shuttle motion accomplished by linear motors instead of force tubes

Figure 6.52 *(continued)*

the forward end of the screw. The extruder die block is closed by a valve, so the accumulating melt forces the rotating screw back against a controlled resistance until sufficient melt has accumulated to make the next parison shot (this action identifies the reciprocating machine action). At this point, usually screw rotation stops.

The second step is the parison extrusion phase. The extruder die block valve is opened and the screw performs the action of a ram by moving forward in the axial direction without rotating. This forces the accumulated melt at the forward end of the screw through the parison head, where it is

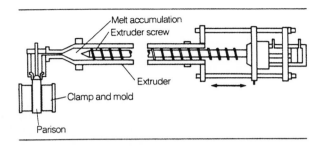

Figure 6.53 Example of a reciprocating screw intermittent extrusion blow-molding machine

extruded at a relatively high flow rate. In practice, the flow rates are limited by the onset of shear phenomena, such as sharkskin and melt fracture.

Thousands of conventional reciprocating injection-molding machines have been built and used worldwide to fabricate blow-molded containers of all sizes. These machines used conventional horizontal-injection molding machines with appropriate continuous operating dies delivering the parisons between all kinds of molds (Fig. 6.54). Mold designs range from simple to those that have all kinds of actions for blow molding complicated products. Decades ago, the IMPCO Machinery Division of Ingersoll-Rand was probably the world leader in manufacturing these types of machines.

Accumulator An intermittent accumulator extrusion blow-molding machine uses a conventional, axially fixed continuously operating extruder to prepare the melt. The accumulator is a heated reservoir where the melt is temporarily stored in the intervals between parison extrusions (similar action of a two-stage injection-molding machine as reviewed in chapter 4) (Figs. 6.55 to 6.57).

The parison is extruded by a ram action on the stored melt. This means that the melt volume in the accumulator fluctuates in a cyclic manner. It is substantially empty immediately after the parison is ram extruded and then it gradually is refilled with melt as the extruded parison enters its blowing phase. The accumulator reaches a maximum fill, based on the parison requirement, immediately after the blown parison leaves the mold.

Accumulators introduce a number of necessary undesirable conditions. They increase the heat history of the melt, and they involve valves and complex flow paths that offer flow resistance and may also lead to difficulties with cleaning and even degradation of the plastic. Accumulators are designed on the FIFO principle (the first material in must be the first out).

Blow-molding accumulators use one of two systems. The separate accumulator is a heated chamber that is an integral part of the blow-molding machine. The capacity of such a chamber can be very large and can, if necessary, be served by several extruders. Figures 6.58 and 6.59 show the other type of accumulator that is built into the parison head. Figure 6.59 call-outs are as follow: 1 = housing, 2 = torpedo, 3 = cardioid groove, 4 = guide legs, 5 = guide ring, 6 = merging zone, 7 = annular ram, 8 = accumulator melt chamber, 9 = die retainer ring, 10 = die, 11 = axially adjustable mandrel, and 12 = axial bore. Other accumulator designs are shown in Figures 6.60 to 6.63.

This type of accumulator takes the form of an annular ram that surrounds the mandrel and torpedo. The construction of the parison head becomes quite complicated and there is the potential for melt leakage and degradation between the moving parts, so design, engineering, and manufacture must be of high standards. First-in-first-out (FIFO) principles are difficult to apply.

Figure 6.54 Series of conventional horizontal injection-molding machines with appropriate blow-molding dies

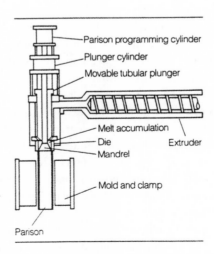

Figure 6.55 Example of an intermittent accumulator head extrusion blow-molding machine

INJECTION BLOW MOLDING

The injection blow-molding (IBM) process uses an injection-molding machine rather than extruder to produce the precursor (chapter 4). This precursor is called a preform rather than a parison as in extrusion blow molding. A major advantage of IBM compared to EBM is that the preform shape can be designed to obtain a uniform (or whatever is desired) wall thickness when blow molded (Fig. 6.64).

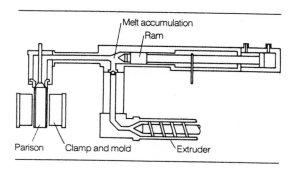

Figure 6.56 Example of an intermittent ram-accumulator extrusion blow-molding machine

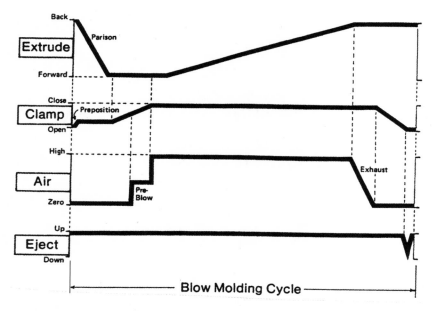

Figure 6.57 Example of the extrusion blow-molding cycle with an accumulator

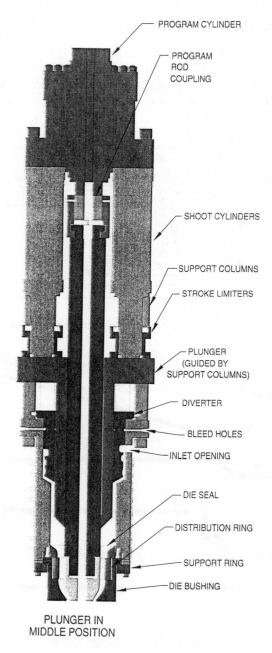

Figure 6.58 Schematic of an assembled intermittent accumulator parison head (courtesy of Graham Machinery Group)

Figure 6.58 *(continued)*

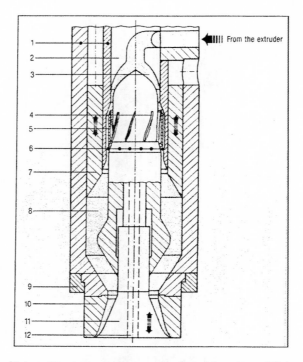

Figure 6.59 Example of intermittent accumulator parison head (courtesy of Bekum)

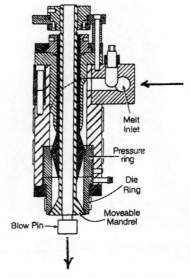

Figure 6.60 Example of intermittent accumulator parison head with a calibrated neck finish

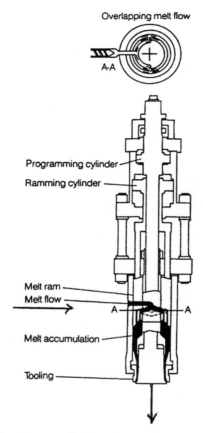

Figure 6.61 Example of intermittent accumulator parison head with overflow melts in the parison to eliminate weld lines

The process consists of blowing a molten thermoplastic against the inside walls of a female mold cavity and chilling it to a rigid solid product. The molding cycle for IBM is shown in Figure 6.65.

The injection blow-molding equipment includes an integral injection component and a multi-impression mold assembly in which mold cores are typically installed on a rotary table. The cores also serve as blowing pins that index in 120° increments between injection, blowing, and ejection stations (Fig. 6.66).

The first station usually has multiple preform injection molds where preforms are formed over core pins. The preforms have hemispherical closed ends (resembling laboratory test tubes). The other ends have open bores, formed by the core pins.

External features, for example the neck flange and threads of a screw-top container, can be produced by injection molding. With the preform still hot and plastic, the mold of the injection machine is opened and the preforms, still on the core pins, are rotated to the second station, which serves as the blowing station. In this position, the preforms are contained by the blow molds with blow molding initiated by introducing air which is blown through the core pins. Next, the blow mold is opened and the finished items, which remain on the core pins, are rotated to an ejection station where they are removed (Fig. 6.67).

In Figure 6.68, the machine has eight multiple core pins in each of the three stations so that after the initial start-up, the three stations operate simultaneously. Most machines in operation

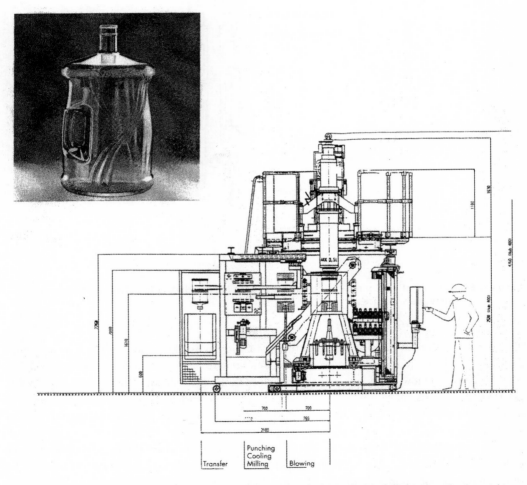

Figure 6.62 Schematic of an EBM with an intermittent accumulator that is fully automatic; insert is an example of a 20-liter (5-gallon) PC plastic bottle fabricated in this machine (courtesy of SIG Blowtec 2-20/30 of SIG Plastics)

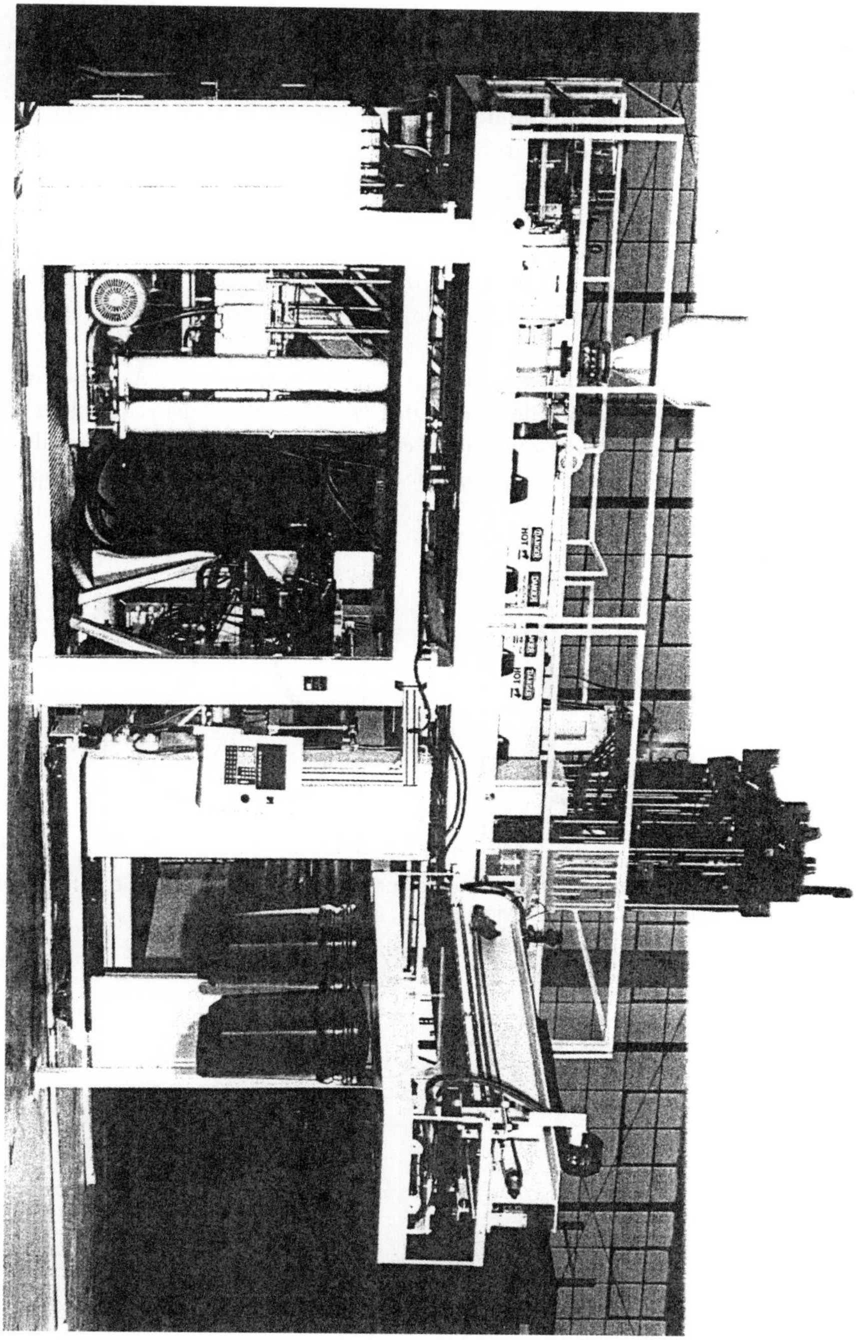

Figure 6.63 Intermittent extrusion blow-molding machine with accumulator molding large tanks (courtesy of Graham Machinery Group)

have many more core pins so that output can be significantly increased. In addition to the three-station IBM machines, there are those that have more stations, with the usual being a four-station machine (Fig. 6.68). The extra station (shown in this figure) is where the preblown preform obtains improved final shape and dimensions and provides a more exact temperature profile condition prior to being blown into its final shape. Other uses for the extra station would include in-mold labeling.

Conventional injection-molding machines with applicable molds have been used to blow mold containers. An example is shown in Figure 6.69 where a shuttle mold is used. After the injection machine prepares the preforms on the cores, the mold opens and is shuttled upward to a blowing station. After containers are ejected from their pins, the mold shuttles back between the machine platens to repeat the cycle.

The injection blow-molding process has a number of advantages. As reviewed, the preform can be injection molded in a profiled shape that corresponds to the requirements of the blow-molded form. The neck form is molded in its entirety at the injection stage, resulting in a quality and precision that is superior to a blow-molded neck. There is no pinch-off scrap to be removed

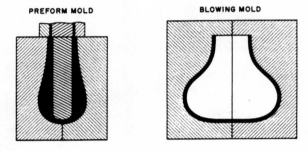

Figure 6.64 Left view shows an injection-molded preform designed to obtain a uniform wall thickness when blow molded (right view)

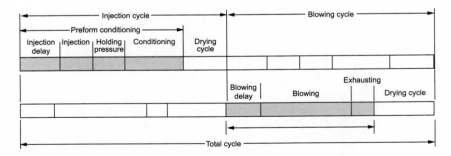

Figure 6.65 Example of the injection blow-molding cycle

and recycled and no bottom seam. There are some drawbacks. The process is difficult to use with high melt viscosity materials. View strips and multiwall constructions are impractical. Integral hollow handles to date are not practical; however, solid integral handles can be included during fabrication. Examples of solid handles are shown in Figures 6.70 and 6.71.

A patent was issued during 1913 (Fig. 6.70) that produced an integral solid handle during unstretched (or stretched) injection blow molding of a bottle. French patent #1,192,475 was issued to the Italian company Manifattura Ceramica Pozzi SpA. The figure shows (a) precision-molded neck that includes the plastic solid handle, (b) preform core and blow pin, (b) basic water-cooled bottle female mold, and (d) injection nozzle of the injection-molding machine.

STRETCH BLOW MOLDING

Stretch blow molding may be performed by the injection or extrusion processes (541). Stretching during blow molding produces biaxial orientation in the blown article (chapters 3 and 5). Conventional blow molding imparts a degree of circumferential orientation, caused by the expansion of the parison into the mold cavity, but there is little or no axial expansion and, correspondingly, no axial orientation. Stretch blow molding provides for axial orientation by stretching the preform or parison axially before or during blowing. This is normally accomplished by means of a stretch rod that is advanced axially inside the preform or parison at a controlled rate. Another axial-stretching system literally clamps the ends and stretches them apart.

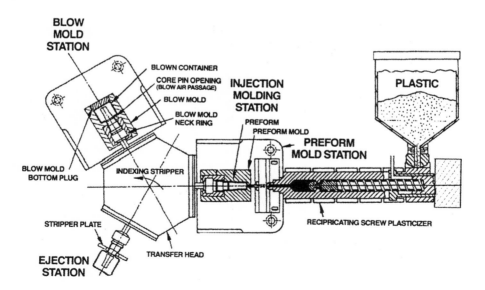

Figure 6.66 Three-station injection blow-molding system

Figure 6.67 Example of ejecting blown containers using a stripper plate

When compared to nonstretch blow molding, stretch blow molding technology provides significant advantages performance- and costwise. The oriented stretch blow types are characterized by factors that include increased strength and rigidity, increased resistance to burst pressure, better transparency and gloss, and reduced permeability. The improved strength can result in savings on additives, such as impact modifiers. Specially developed nucleated or clarified random copolymer grades produce a blown bottle of glasslike transparency. Very significant is that they can be made lighter and cheaper, offsetting the higher capital cost of equipment. The lighter weight stretched containers result in reduced handling, transport, and warehousing costs. Stretch blow processes can produce containers with solid integral handles (Fig. 6.70).

INJECTION STRETCH BLOW MOLDING

ISBM produces biaxial orientation that provides increased tensile strength (top load); less permeation due to tighter alignment of the molecules; improved drop impact and clarity; and a lightweight container. The injection-molded preform is usually short and thick-walled relative to the finished blown article. The neck profile complete with screw thread is entirely formed by injection molding and is generally not modified by the blowing process. The other end of the preform is closed and typically dome shaped. The design and precision of the preform have critical influences on the degrees of orientation and quality of the blown articles. The stretch blow-molding processes continue to be in widespread use worldwide for producing injection blow-molded polyethylene terephthalate (PET) carbonated drink bottles and other plastics, such as PVCs. The injection-molded preform may be converted to a blow molding either by the single-stage or the two-stage process.

Not all thermoplastics can be oriented. The major thermoplastics used are polyethylene terephthalate (PET), polyacrylonitrile (PAN), polyvinyl chloride (PVC), and polypropylene (PP). PET is by far the largest volume material, followed by PVC, PP, and PAN. The amorphous materials with a wide range of thermoplasticity, such as PETs, are easier to stretch blow than the partially crystalline types, such as PPs (chapter 1). Approximate melt and stretch temperatures to yield maximum container properties are shown in Table 6.9.

Single-Stage In the single-stage (or one-step) injection stretch blow-molding process, the preform injection-molding step is integrated with the stretch blow machinery. The machines are generally arranged for rotary operations so that the preforms pass directly from the injection-molding station to a thermal-conditioning station and then to a stretch blow-molding station (Figs. 6.72 to 6.76). The Figure 6.72 call-outs are as follows: 1 = preform injection-molding station, 2 = thermal-conditioning station, 3 = stretch blow-molding station, and 4 = ejection station.

Single-stage equipment is capable of processing PVC, PET, and PP. Once the parison is formed (either injection or extruded), it passes through conditioning stations that bring it to the proper

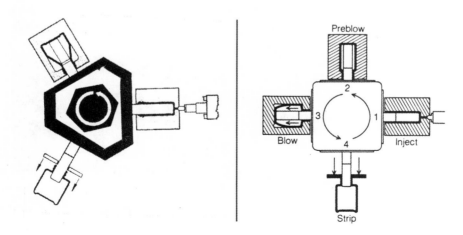

Figure 6.68 Examples of three-station and four-station injection blow-molding machines

orientation temperature. The single-stage system allows the process to proceed from raw material to finished product in one machine, but since tooling cannot be easily changed, the process is best suited for dedicated applications and low volumes.

Many oriented PET containers are produced on single-stage machines. The preform is first injection molded, then transferred to a temperature-conditioning station, then to the blow-molding operation where it is stretch blown into a bottle, and finally to an eject station.

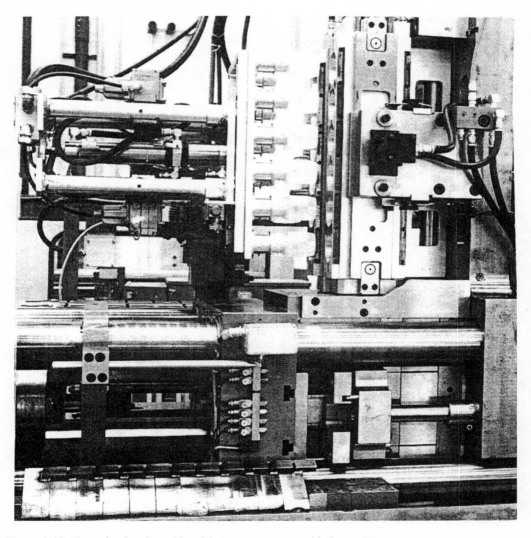

Figure 6.69 View of a shuttle mold to fabricate injection-molded containers

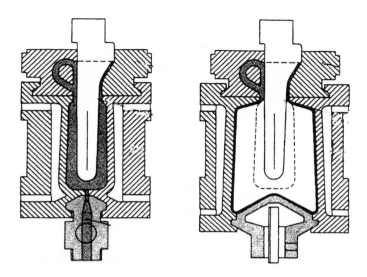

Figure 6.70 Schematic of injection blow mold with a solid handle

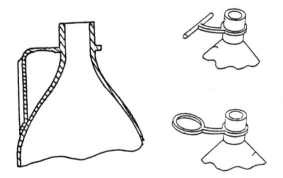

Figure 6.71 Simple handles (ring, strap, etc.) can be molded with blow-molded bottles and other products

Material	Melt, °C.	Stretch, °C.
PET	280	107
PVC	180	120
PAN	210	120
PP	240	160

Table 6.9 Examples of plastic melt and stretch temperatures

This process can be used to produce wide-mouthed jars as well as necked bottles, but it is less suitable for high-speed production than the two-stage process. The injection-molded preform is cooled rapidly in the mold to about 90°C (Fig. 6.77), and the core pin is extracted as soon as the surface skin has cooled sufficiently to make the preform physically stable. The pin has a generous draft angle to make this easier.

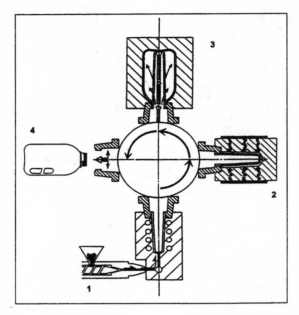

Figure 6.72 Single-stage injection stretch-blow process

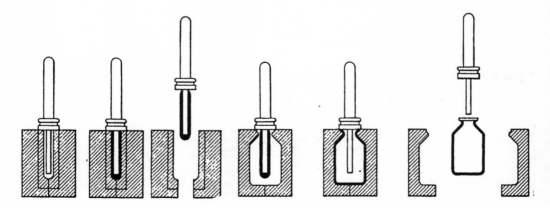

Figure 6.73 Schematic of the steps taken for injection stretch blow molding

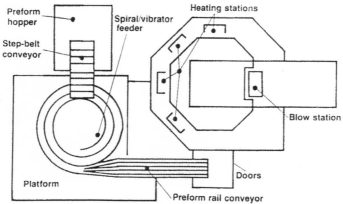

Figure 6.74 Schematic and internal view of a fast-operating reheat preform for stretched IBM (courtesy of SIG Plastics International)

Thermal conditioning is the most critical part of this process. The optimum stretch-blow temperature span can be very narrow, ranging from within about 10°C. Uniform heating of the order of ±1°C is important. Thermal-conditioning stations include up to thirty-six independently controlled heat zones, typically using a combination of infrared radiant external heaters and hot air for internal heating.

The thermally conditioned preform is transferred to a blow mold and stretched and oriented axially by an internal stretch rod or mechanical grip stretching, either immediately prior to or simultaneously with the blowing operation that provides radial stretch and orientation (Figs. 6.78 and 6.79). Blowing pressures range up to about 40 bar. The blow mold temperature is relatively high at 35° to 65°C in order to minimize strain in the bottle. For a given bottle size, the degree of orientation is determined principally by the parison length and diameter. Stretch ratios are relatively high. In the wall thickness of the bottle body, the amount may be as high as 15:1. Axial stretch is about 4:1; diametrical stretch ranges about 3.5:1.

Two-Stage This process uses two different operating machines (Fig. 6.80). The two-stage (two-step) or reheat injection stretch blow-molding process completely separates the preform injection

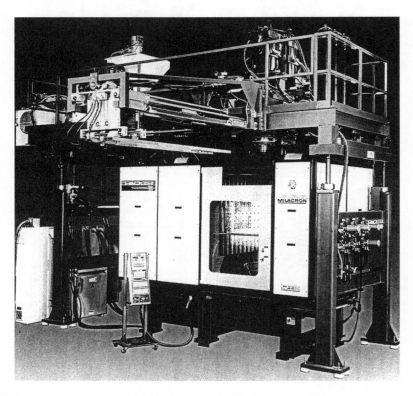

Figure 6.75 Easy-to-operate and control in-line stretch IBM (courtesy of Milacron)

molding operation from the blow-molding operation. The two operations may be carried out in different locations by different manufacturers, with a substantial time interval between them. The process lends itself to high rates of production and has the advantage that preform production may be entrusted to a specialist processor (Figs. 6.81 and 6.82).

This process requires two heating operations, resulting in a greater energy use and increased heat history in the plastic. It also requires the need to store and handle preforms. Before being

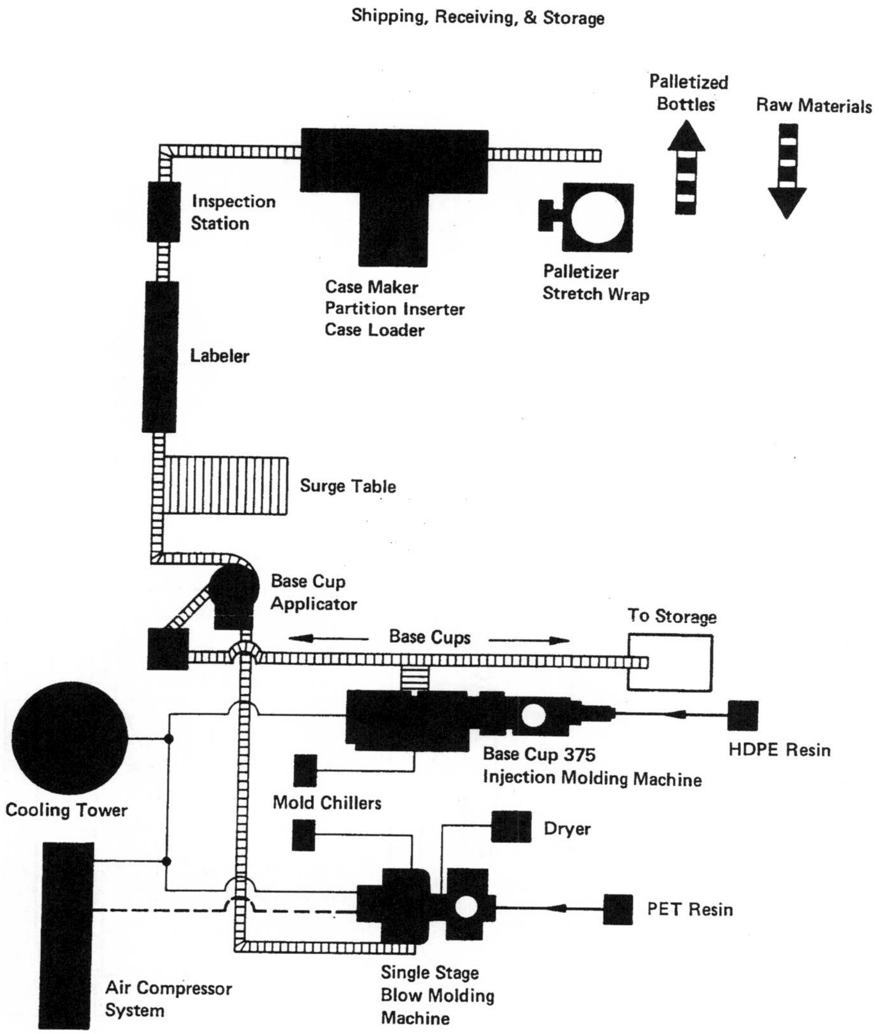

Figure 6.76 Example of a single-stage injection stretch blow-molding production line

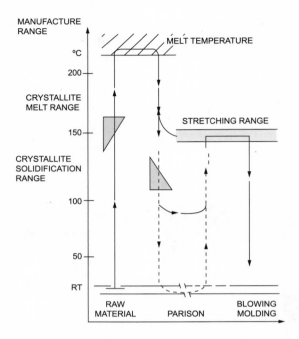

Figure 6.77 Temperature range for stretch blow molding polypropylene

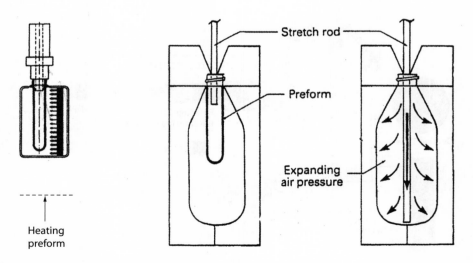

Figure 6.78 Example of stretched injection blow molding using a rod

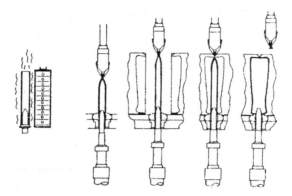

Figure 6.79 Example of stretched injection blow molding by gripping and stretching the preform

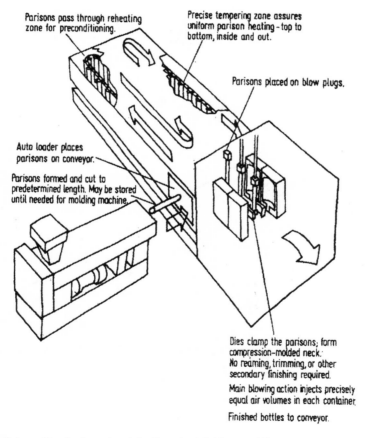

Figure 6.80 Schematic of a two-step injection stretch blow-molding process (courtesy of Milacron)

blown, the preforms are reheated from room temperature and are then thermally conditioned and stretch blown in a similar manner to the single-stage process. The two-stage or reheat stretch-blow machine is usually arranged for rotary high-speed continuous operation. Even with these types of situations, the two-stage process overshadows them with its advantages.

With the two-stage process, processing parameters for both preform manufacturing and bottle blowing can be optimized. A processor does not have to make compromises for preform design and weight, production rates, and bottle quality as done on single-stage equipment. One can either make or buy preforms. If one chooses to make them, one can do so in one or more locations suitable to the market. Both high-output machines and low-output machines are available.

Two-stage extrusion-type machines generally are used to make oriented PP bottles. In a typical process, preforms are extruded, cooled, cut to length, reheated, stretched while the neck finish is being trimmed, and ejected. The two-stage process has been the lowest-cost method to produce

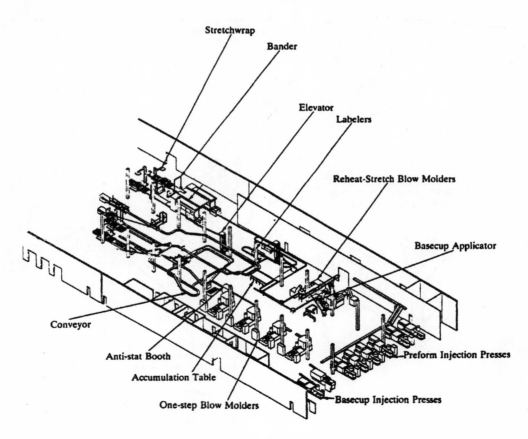

Figure 6.81 Example of a bottling plant using the two-step injection stretch blow-molding process

oriented PET containers. Two-stage stretch blow molding also is being used for production of oriented PVC containers.

The two-stage process, which permits injection molding of the preform and then shipping to blow-molding locations, has allowed companies to become preform producers and to sell to blow-molding producers. Thus companies that wish to enter the market with oriented containers can minimize their capital requirements.

Preform design and its relationship to the final container remain very critical factors. The proper stretch ratios in the axial and hoop directions must be met if the container is to properly package its intended product (Table 6.10).

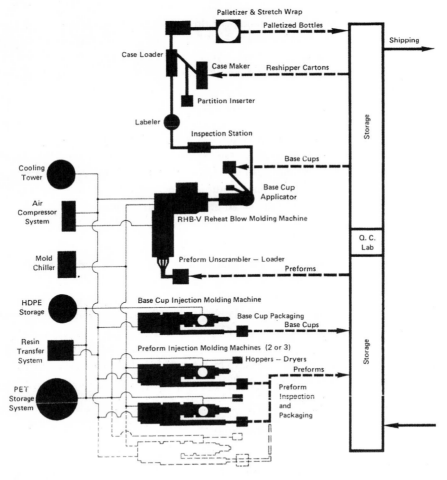

Figure 6.82 Example of a two-stage injection stretch blow-molding production line

Material	Stretch ratios	Orientation temp., °F.
PET	16/1	195-240
PVC	7/1	210-240
PAN	9/1	220-260
PP	6/1	260-280

Table 6.10 Examples of stretch ratios for different plastics

Specifications should be established on the preform, which should be produced with as little molded-in stress as possible. The molded-in stress can be observed by looking at the preform using cross-polarized sheet with a light source. Molded-in stress will cause blowouts, off-center gates, irregular wall distribution, poor top load, and poor drop test results.

Other critical factors in the preform, such as with PET resin, are concentricity (not to exceed 0.005 in TIR), end cap crystallization (diameter at the gate not to exceed 0.25 in), length of the gate (0.125 in maximum), weight of the preform (± 0.25 g) over the desired weight, unmelt in the preform, and air entrapment (bubbles).

Technologies, such as coextrusion and coinjection of preforms, allow PET and other plastics to package foods and other products. Care must be taken to control the process so that the preform when stretch blown will not have microvoids in the container walls and will not delaminate. Heat setting of PET containers during stretch-blow molding is required. The process must be constructed so that a specific percent of crystallinity is imparted to the blown container plus a temperature memory of at least 15°F above the filling temperature of the product. Container design can be used to assist the properties of percent crystallinity and the temperature memory.

SPECIAL MACHINES

Different stretch blow molders are designed and built to meet specific products. For example, Mitsui Machine Technology Inc. is fabricating a new way to make PET cups using a stretch injection blow-molding machine produced by Frontier Inc. of Ueda City, Japan. Frontier's rotary, double-axis draw blow-molding system has two main components: a heating section and a blow-molding unit that shapes a heated PET preform into a cup.

This all-electric machine can produce 240 cups per minute. This blow-forming method provides high vertical draw, which in turn provides a high rate of axis draw, improving thickness distribution and enhancing oriented crystal density. The result is a cup that is 40% stronger than similar thermoformed cups (537).

EXTRUSION STRETCH BLOW MOLDING

Extrusion stretch blow molding is a one-stage or two-stage process using two mold/mandrel sets, where one is for preblow and the other for final blow. An extruded parison is first pinched off and blown conventionally in a relatively small preblow mold to produce a closed-end preform. The preform is then transferred to the final blow mold, where usually an extending stretch rod within the blowing mandrel bears on the closed preform end to stretch it axially. The stretched preform is then blown to impart circumferential stretch. Standard blow-molding machines can be converted for extrusion stretch blow molding. The process is most often used for PVC bottles.

Oriented PVC containers most commonly are made on single-stage, extrusion-type machines. The parison is extruded on either single- or double-head units. Temperature conditioning, stretching, and thread forming are done in a variety of ways depending on the design of the machine. Many of the processes in use are proprietary.

DIP BLOW MOLDING

The dip blow molding process bears some resemblance to injection blow molding in that it is a single-stage process performed with a preform on a core/blow pin (Fig. 6.83). This figure's call-outs are as follows: 1 = accumulator filled, 2 = blow pin inserted, 3 = neck mold filled, 4 = blow pin partially withdrawn, 5 = blow pin fully withdrawn at variable speed, 6 = preform trimmed, 7 = preform clamped in mold, and 8 = preform blow molded.

The difference is in the way the preform is made. The process uses an accumulator cylinder that is fed by an extruder. The cylinder has an injection ram at one end while the other is a free fit over the blow pin. The blow pin is dipped into the melt so that a neck mold on the pin seals the end of the accumulator cylinder. The injection ram is advanced to fill the neck mold; then, the blow pin is withdrawn at a controlled rate so that it is coated with a melt layer extruded through the annular gap between the pin and the accumulator cylinder. The thickness of the coating can be varied or profiled to an extent by varying the speed of the blow pin and the pressure on the injection ram. After trimming, the preform is blow molded in the same manner used for injection blow molding.

The process results in a seam- and flash-free container with a high-quality molded neck. The preform is produced at a lower pressure than that used for injection molding, so the machine can be

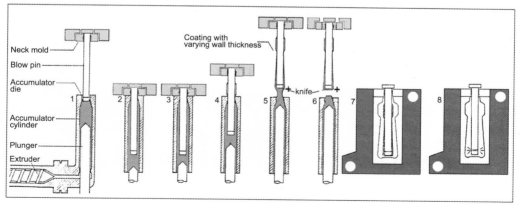

Figure 6.83 Stages in the dip blow-molding process

more lightly constructed and at lower cost. The preform is formed under relatively low stress. Like injection blow molding, the process is best suited to the production of smaller containers.

Multibloc Blow Molding

The multibloc process is used for high-volume blow molding of very small containers, such as pharmaceutical vials and whiskey bottles. A multicavity mold is used with an extruded parison whose circumference approaches twice the total width of the closely spaced cavities. Before the mold closes, the parison is stretched and semi-flattened laterally so that it extends across the full width of the cavities (Fig. 6.84). The process is usually combined with blow/fill/seal techniques.

Other Blow-Molding Processes

Coextrusion and Coinjection Coextrusion and coinjection (or multilayer processes) are essential techniques in the production of high-performance blow-molded products (chapters 4 and 5). The parison or preform is coextruded with a number of different layers, each of which contributes an important property to the finished product. Increasingly, a midlayer may consist of recycled material that is encapsulated between inner and outer layers of virgin plastics.

Blow moldings commonly include from two to seven layers, although more are also used. The construction usually includes one or more barrier layers. These are plastics that have particular resistance to the transmission of water vapor or gases, such as oxygen or carbon dioxide. Examples

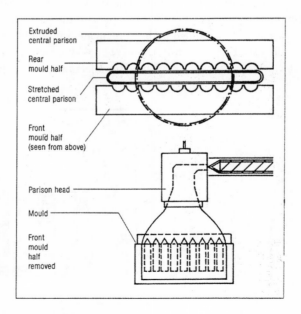

Figure 6.84 Multibloc blow-molding process

are ethylene vinyl alcohols (EVOHs), nylons, and polyvinylidene chlorides (PVDCs). Their presence greatly enhances the performance of the blow molding as a package for foodstuffs, beverages, and other critical products. Each barrier material is deficient in some respect, such as price, mechanical strength, and moisture resistance, and thus are not used as a material of sole construction for blow molding. Their use is in thin layers shielded by other more robust and economical body plastics.

Two considerations lead to high-layer counts in coextruded or coinjection blow-molded products. The barrier materials may be incompatible with body materials, such as polyethylenes and polypropylenes, so intermediate or tie layers of a mutually compatible adhesive must be included to bring about a bond between the layers. Some barrier layers are so deficient in other respects that they need protective layers before they can be used effectively.

As an example, ethylene vinyl alcohol has outstanding resistance to oxygen transmission but has high rates of moisture absorption and water vapor transmission, both of which damage its oxygen barrier performance. The plastic needs both protective layers and adhesive layers, which contribute five layers whenever used. An example using a tie layer with ethylene vinyl alcohol is an ethylene-vinyl acetate-maleic anhydride terpolymer. As shown in Figure 6.85, a typical coextruded polypropylene bottle will include body, regrind, barrier, and tie layers. This figure's call-outs are as follows: 1 = inner body layer of polypropylene, 2 = reground material, 3 = tie layer, 4 = ethylene vinyl alcohol barrier layer, 5 = tie layer, and 6 = outer body layer of polypropylene. Figures 6.86 and 6.87 show examples of coinjection blow-molded bottles.

Such containers are fabricated by performing conventional extrusion or injection blow-molding operations on a multilayer parison or preform. A coextrusion parison head, served by a separate extruder for each distinct component material, is used to produce the parison. The barrier and tie layers are usually very thin, so the flow engineering in the head is critical in order to preserve the integrity of the layers. For this reason, the various melt streams are merged as close to the die exit as possible, even though this complicates die-head construction (Figs. 6.88 and 6.89). Other disturbing influences, such as parison profiling and intermittent extrusion, are often avoided. If

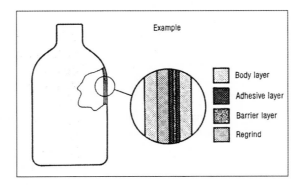

Figure 6.85 Example of a six-layer coextruded blow-molded bottle

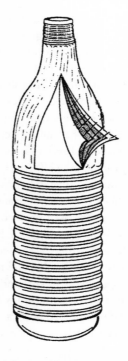

Figure 6.86 Example of a five-layer coinjection blow-molded bottle

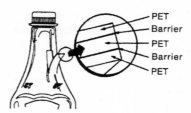

Figure 6.87 Example of a five-layer coinjection blow-molded ketchup bottle

parison profiling is required, the mechanical complication of the parison head is such that axial movement of the die rather than the mandrel usually occurs.

Multilayer blow-molded bottle markets include foods and industrial and household chemicals. The food market usually requires a minimum of five layers and preferably six where regrind can be used. An example of a layer structure for food containers consists of PP/adhesive/EVOH/adhesive/PP. This produces a container that can be hot filled (Fig. 6.90), is relatively inexpensive, has decent

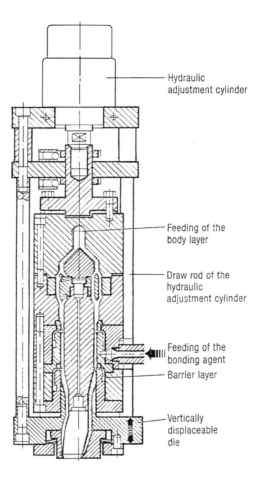

Figure 6.88 Example of a three-layer coextrusion parison blow-molded head with die profiling

contact clarity, and will have a shelf life of approximately one year without refrigeration. The industrial and household chemical markets can be met with three-layer combinations where barriers, such as nylons or acrylonitriles, are placed on the inside and the inexpensive HDPE resin on the outside.

The machinery used primarily for blow molding multilayer containers is extrusion blow-molding equipment of a proprietary type. These machines are the rotary wheels, where multiple extruders feed a single parison head. From four to eighteen cavities may be used on these rotary wheels. These machines have the advantage of using only a single parison, since multilayer structures are difficult to split for multiple parisons.

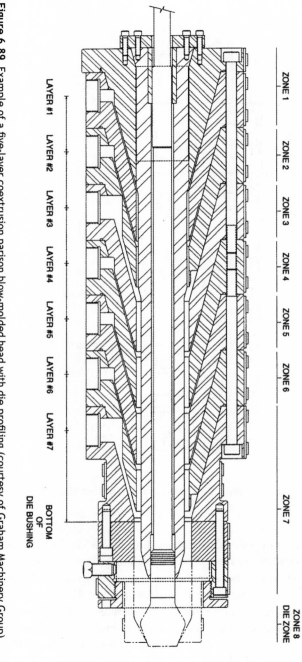

Figure 6.89 Example of a five-layer coextrusion parison blow-molded head with die profiling (courtesy of Graham Machinery Group)

The output that can be achieved on rotary wheels is at least up to 5,000 per hour. The disadvantages are that neck finishing must be done in a secondary operation and there is generally more scrap generated since most wheels do not use cut parisons. There is flash between the molds that must be reused in the process. The shuttle-type platen machines can be equipped quite easily with extruders and a multilayer head to produce bottles with calibrated neck finishes and a minimum amount of scrap. The disadvantage of a platen shuttle-type machine is the output requirements since the multilayers are difficult to split in flow to produce more than two parisons. The output that can be achieved on platen-type machines, with two parisons, is up to 1,200 per hour. The tie layers used are extrudable adhesives and work well in small extruders. The key factor in the head design is to keep the adhesive layers and the barrier layer at a minimum since these materials are expensive.

As reviewed, the other method for producing multilayer bottles is by coinjection blow molding where a system of two or more layers produces bottles. With the use of a barrier layer, it is encapsulated between other structured layers up to the preform neck ring, which ensures barrier integrity and prevents air entrapment. The injection-blow multilayer system is primarily used to fabricate PET soft drink bottles where the barrier layer achieves a longer carbonation retention.

Sequential Extrusion Sequential extrusion blow molding is a special multimaterial technique used for the production of specially designed products. The different plastics are chosen typically to

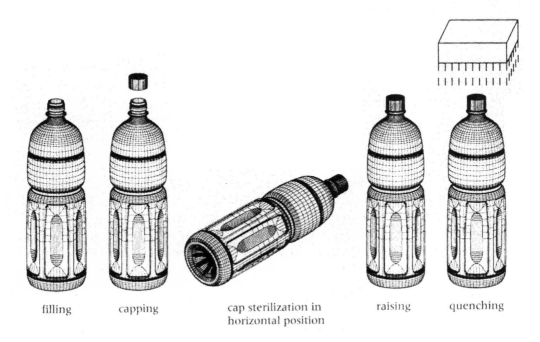

Figure 6.90 Example of hot-filling PET bottle at 80° to 95°C (courtesy of SIG Plastics International)

contribute complementary mechanical properties and are present in distinct sequential zones in the finished part. Normally, two materials are used but three or more are also used. Separate external ram accumulators for each material serve the die head. These are operated sequentially, typically in A-B-A sequence, to produce a parison with three distinct material zones in axial succession. The parison is subsequently blow molded by normal techniques.

An example for sequentially blow molding polypropylene is an automotive air duct in which a central flexible zone (Figs. 6.16 and 6.91) joins rigid end sections. The flexible zone allows for installation mismatches, accommodates thermal expansion, and damps vibration noise. The rigid portions allow for direct connections to other mechanical elements in the assembly.

Blow/Fill/Seal The blow/fill/seal process is a complete packaging technique that integrates the extrusion or injection blow-molding and container-filling steps. This provides for aseptic filling of the hot as-blown container and is used for pharmaceutical, food, and cosmetic products. The process employs a two-part mold in which the container body mold cavity blocks are separate from the neck-forming members (Fig. 6.92).

The body mold closes on the parison that is blown normally by a neck-calibrating blow pin. Immediately, with the mold still closed, the liquid contents are injected through the pin. The pin is then withdrawn and the neck is formed and sealed under vacuum by the neck-forming members. Both mold parts then open to eject a filled and sealed container. Small containers may be formed entirely by vacuum rather than blowing.

Blow/fill/seal machines are specially adapted for clean and sterile working. The parison knife runs at temperatures up to 400°C to ensure a sterile cut. Blowing air is generated by an oil-free compressor and is filtered down to 0.2 μm in sterile filters. The filling region of the machine is swept by sterile air and the entire machine is housed in clean room conditions. The normal melt temperature for extrusion is sufficient to produce a sterile parison. Complete blow/fill/seal cycle times typically range from nine to twenty seconds. Different plastics are used to meet different end-item performances. For example, polypropylene is a leading material for blow/fill/seal applications, generally in the form of high-purity food-approved and other grades free of potentially harmful additives.

Blow Molding 3-D Because extrusion blow molding is performed on a cylindrical parison, the conventional process is not well suited to the production of products with complex forms that deviate substantially from the parison axis. Such forms can be produced by conventional blow molding equipment but only by using a parison that in its form blankets the complex mold cavity. This 3-D process in the past usually developed an excessive amount of pinch-off scrap. During the past few decades, developments in parison-handling robot equipment and in blow-mold designs made it possible to manipulate relatively small parisons into complex mold cavities. The result is a blow molding largely free of flash and scrap and offering considerable process savings. There are many such techniques, some of them proprietary property, and they are collectively known as 3-D blow molding. An example is shown in Figure 6.93. This figure shows the steps involved in fabricating a complex 3-D shape: the mold begins to close on a prepinched and preblown parison, next is to mold fully closed with the core forward and the side walls hinged, and the final step is to open the mold and retract the core.

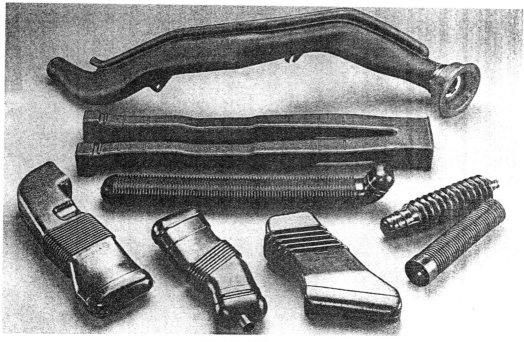

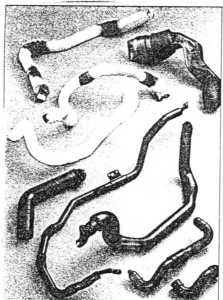

Figure 6.91 Examples of different shaped sequential extrusion blow-molding products

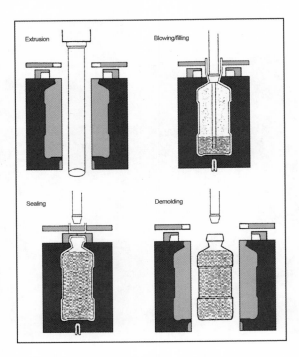

Figure 6.92 Example of container-filling steps in the blow/fill/seal extrusion blow-molding process

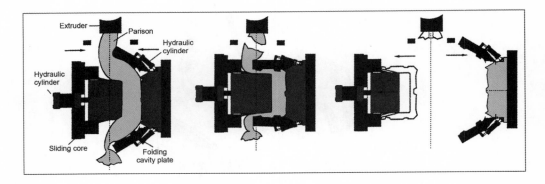

Figure 6.93 Example of a 3-D extrusion blow molding process (courtesy of Placo)

Other 3-D blow-molding methods are shown in Figures 6.94 to 6.98.

BLOW MOLDING WITH ROTATION

Injection molding with rotation (MWR) is an example of processing at lower temperatures, pressures, and so on. It is also called injection spin molding or injection stretched molding. This blow-molding process combines injection molding and injection blow molding (and performs like injection blow molding, as reviewed) but has an additional step of melt orientation (Dow Chemical patent). The equipment used is what is commercially available for IM except the mold is modified so that either the core pin or outside cavity rotates. The rotated melt on its preform pin is transferred to a blow mold (Fig. 6.99). The end product can come directly from the IMM mold or be a result of two-stage fabrication: making a parison and blow molding the parison.

This technology is most effective when employed with an article that has a polar axis of symmetry and reasonably uniform wall thickness as well as dimensional specifications and part-to-part trueness that meets market acceptance. The MWR process requires no sacrifice of either cycle time or surface finish. Both laboratory and past commercial runs identify good potentials for reducing cycle times, for reducing the amounts of plastic required or improving properties with the same amount of plastic (or both), and for substituting less expensive plastics while achieving adequate properties in the fabricated products.

During fabrication using the MWR process, two forces act on the plastic: injection (longitudinal) and rotation (hoop). The targeted balanced orientation is a result of those forces. As the product's walls cool, additional high-magnitude cross-laminated orientation is frozen in and develops throughout the walls. Orientation on molecular planes occurs as each layer cools after injection. This orientation can change direction and magnitude as a function of wall thickness. The result is analogous to plywood or reinforced plastics (chapter 15) and the strength improvements are as dramatic. In the MWR process, there are an infinite number of microscopic layers, each of which has its own controlled direction of orientation. Using appropriate processing conditions, the magnitude and direction of the orientation and strength properties can be varied and controlled throughout the wall thickness.

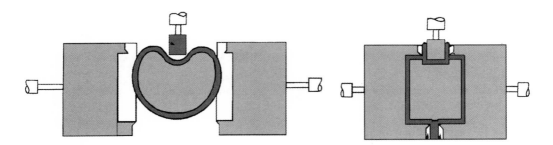

Figure 6.94 Examples of multiple side action 3-D extrusion blow-molding molds

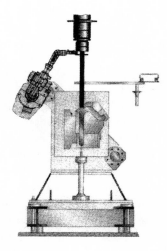

Phase 1
- Extrusion of parison and gripping
- Cutting of parison

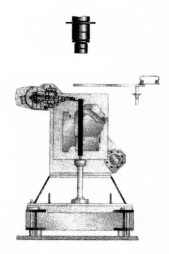

Phase 2
- Lowering of parison and closing of 1st part of mold

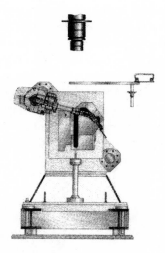

Phase 3
- Insertion of parison
- Closing of remaining mold parts
- Blow molding of article

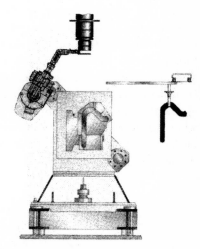

Phase 4
- Opening of blow mold
- Removal of article

Figure 6.95 Example of six-axis robotic control to manipulate a parison in a 3-D mold cavity to extrusion blow mold products (courtesy of SIG Plastics International)

MOLD

Essentially, the blow mold consists of two halves, each containing cavities that, when the mold is closed, define the exterior shape of the blow molding. Because the process produces a hollow article, there are no cores to define the inner shape. The blow-molding process is carried out at relatively low pressures, so blow molds can be more lightly engineered than injection molds and

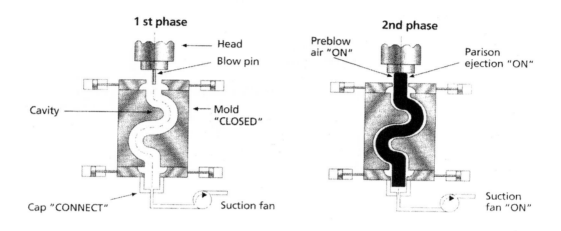

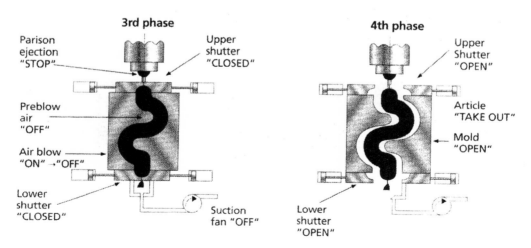

Figure 6.96 Example of a suction 3-D extrusion blow-molding process (courtesy of SIG Plastics International)

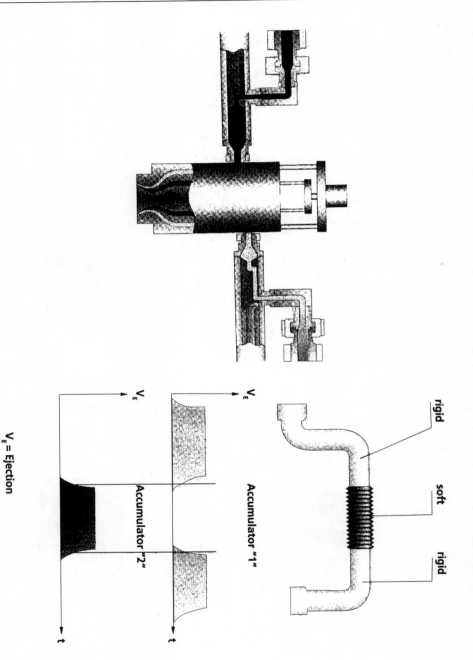

Figure 6.97 Example of sequential 3-D coextrusion blow-molding machine (courtesy of SIG Plastics International)

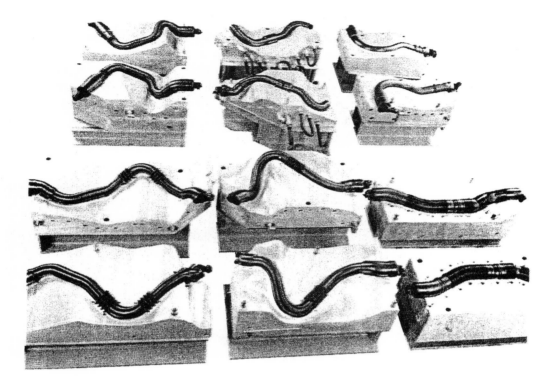

Figure 6.98 Examples of 3-D extrusion blow-molded products in their mold cavities (courtesy of SIG Plastics International)

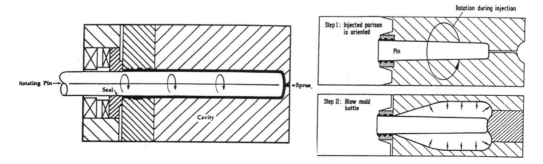

Figure 6.99 Schematic for molding with rotation using a two-stage blow-molding procedure

are correspondingly cheaper and quicker to manufacture (chapter 17). Single-cavity molds are used when working with a single parison. When the usual multiple parisons are in use, a number of single-cavity molds may be mounted on the machine platen or the requirement may be met by a Multiple-cavity mold. Mold details and actions will vary considerably according to the geometry of the product and the blow-molding process in use. Even though the following review concentrates on extrusion blow molding, the information can also be applied to injection blow molding. Table 6.11 provides an introduction to a mold design checklist.

BASIC FEATURES

As shown in Figure 6.100, the usual mold consists of two halves that meet on a plane known as the parting line. The plane is chosen so that neither cavity half presents an undercut in the direction of the mold opening. For most bottle designs, this requirement presents little or no difficulty. Figure 6.101 is one of the many hundreds of different blow-molding molds used to fabricate different products.

For products of asymmetrical cross-sections, the parting line is placed in the direction of the greater dimension (Figs. 6.102 and 6.103). Guide pillars/pins and bushings to ensure that there is

1.	Was latest issue part drawing used?
2.	Will mold fit press for which intended?
3.	Is daylight of press sufficient for travel and ejection?
4.	Do guide pins enter before any part of mold?
5.	Can mold be assembled and dis-assembled easily?
6.	Has allowable draft been indicated?
7.	Is plastic material and shrinkage factor specified?
8.	Are mold plates heavy enough?
9.	Are waterlines, air lines, thermo-couple holes or cartridge holes shown and specified?
10.	Is ejector travel sufficient?
11.	Is the material type for mold parts specified?
12.	Do loose mold parts fit one way only? (Make fool proof.)
13.	Will molded part stay on ejector side of mold?
14.	Can molded part be ejected properly?
15.	Have trade marks and cavity numbers been specified?
16.	Has engraving been specified?
17.	Has mold identification been specified?
18.	Has plating or special finish been specified?
19.	Is there provision for clamping mold in press?
20.	And others . . .

Table 6.11 Mold design checklist

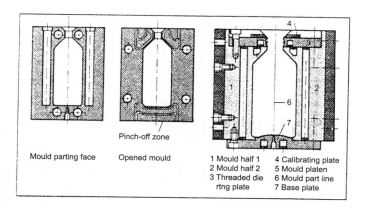

Figure 6.100 Example of an extrusion blow mold

no mismatch between the cavities align the two mold halves. With extrusion blow molding, the parison passes across the mold in the axis of the cavity and is pinched and compressed between the faces of the closing mold at the neck and base regions of the cavity. These are known as the pinch-off zones. Separate inserted mold blocks typically form the base and neck regions of the mold. The mold includes channels for the circulation of cooling water. With injection blow molding the preform only has a pinch-off at the neck.

Materials of Construction

The key factors influencing the choice of materials for a blow mold, as with molds for other processes, are durability, thermal conductivity, finished cost, and surface finish. The main cavity body for blow molds is usually produced in machined aluminum alloy, cast zinc, or another nonferrous alloy. For long production runs, steel or cast beryllium copper may be preferred (Table 6.12; chapter 17). The pinch-off zones are subject to compressive stress and wear, and these are usually furnished as inserts of alloy steel or hard beryllium copper.

Pinch-Off Zone

In extrusion blow molding, the pinch-off zone performs two functions. It must weld the parison to make a closed vessel that will contain blowing air and it must leave pinched-off waste material in a condition to be removed easily from the blown product. To accomplish this, the pinch-off zone has a three-stage profile: pinch-off edge, pressing area, and flash chamber (Figs. 6.104 and 6.105).

The pinch-off edge is flush with the mold parting line and forms part of the cavity periphery. The edge should be as narrow as possible depending on the material of construction. For example, steel edges can be from 0.3 to 1.5 mm wide for rigid plastics; for softer materials, the dimensions can be 0.8 to 2.5 mm. The pressing area is recessed into the mold parting face by an amount depending on the parison thickness. Its function is to displace melt into the weld area to ensure a strong weld that is substantially equal in thickness to the blown wall thickness at the same point on the parison axis.

The pressing area depth is typically determined by trial and error and is often given a ribbed surface to improve cooling of the waste material by increasing the area available for heat transfer.

Figure 6.101 Blow-molded corrugated bellow part between its mold halves

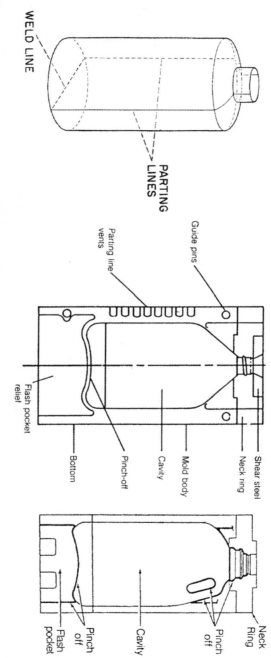

Figure 6.102 Examples of parting line locations and other parts of a mold

The flash chamber is recessed to a greater depth than the pressing area, typically to about the thickness of the parison. If it is made too deep, heat transfer will suffer and the parison scrap will cool too slowly. Its function is to limit the extent of the pressing area and so limit the force opposing the mold closing. Angling the riser face by 30° to 45° to the parting face will strengthen the steps between the different levels.

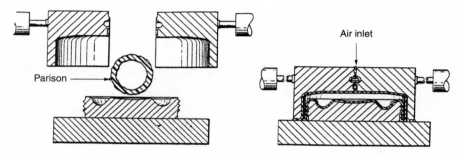

Figure 6.103 Example of a three-part mold to fabricate a complex threaded lid

Material	Hardness[1]	Tensile strength		Thermal conductivity Btu/in./ft² h F
		psi	MPa	
Aluminum				
A346	BHN-80	36.975	255	1.047
6061	BHN-95	39.875	275	1.165
7075	BHN-150	66.700	460	905
Beryllium copper 23 and 165	RC-30 (BHN-285)	134.850	930	728
Steel				
0-1 and A-2	RC 52-60 (BHN-530-650)	290.000	2.000	243
P-20	R-32 (BHN-298)	145.000	1.000	257

BHN = Brinell and RC = Rockwell hardness (C scale).

Material	Wear resistance	Ability to be cast	Ability to be repaired	Density lb/cum (g/cm³)	Ability to be machined and polished
Aluminum	poor	fair	good	0.097 (2.699)	excellent
Beryllium-copper	excellent	good	good	0.129–0.316 (3.589–8.793)	fair
Cast iron	good	good	good	0.24 (7.6)	good

Table 6.12 Examples of materials used in the construction of blow-molding molds

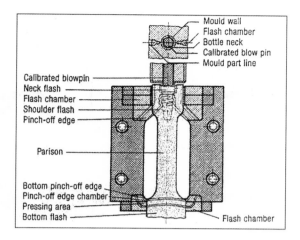

Figure 6.104 Examples of pinch-off zones in an extrusion blow mold

FLASH CONTROL

Flash caused by the pinch-off is an unavoidable evil in extrusion blow molding. The ability to control the adverse effects of the flash is critical to success of the process (543). Pinch-off generates excess material in the form flash that is usually twice the thickness of the part's wall. This thicker plastic cools slower than the blown product. It is subject to fold-over and adheres to the blown product.

Flash imposes costly limits on the efficiency of blow molding. It has potential for significantly extending the molding cycle, primarily by increasing the time needed to cool the thick flash. This cycle increase could approach twice what would normally be required. Removal calls for a postmolding trim step that requires secondary equipment and poses a risk of damaging good parts.

To reduce cycle time, a fabricator has some damaging options, such as ejecting the part before the flash is sufficiently cooled. Because it is still soft and pliable when ejected, it can create other problems, such as a fold over on itself and adhering to adjoining surfaces of the part after ejection of the molding. Flash is also considerably more difficult to handle and trim while hot. In either case, the resultant penalty may be a significant increase in the part reject rate.

By locating cooling lines as close as possible to the flash, heat transfer to the cooling water will reduce cycle time. It is critical to appreciate maximizing the heat transfer as much as possible to the flash area. Keep the water turbulent by operating the water in the proper Reynold's number (chapter 17).

In most bottle molds where flash occurs, it always has a separate insert. That is because the pinch-off area traditionally requires a metal insert (such as QC-7 aluminum) that provides resistance to wear and increased heat-transfer efficiency. When the insert starts to become inefficient, it is replaced with a new insert.

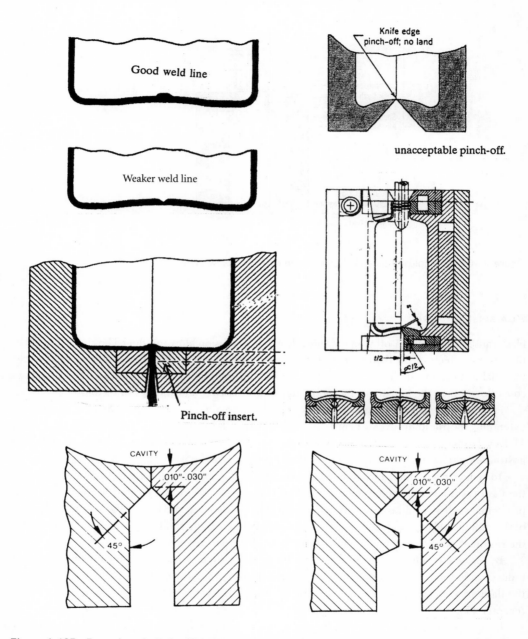

Figure 6.105 Examples of pinch-off designs to meet requirements for different plastics and contours

The cross-section of the insert's parting line in a conventional blow mold traditionally takes the form of a horizontal line, the main reason being that it is easier and less costly for a tool builder to machine in straight lines. However, a trapezoidal cross-section would expand the surface area of the mold in contact with the flash, thereby removing heat more efficiently (Fig. 6.106). A trapezoidal insert cross-section can also increase the flash's stiffness, making the hot flash more resistant to folding over.

BLOWING AND CALIBRATING DEVICE

The blow pin is the means by which blowing air is inserted into the parison through what will become a hole in the finished blow molding. In the case of a bottle (Fig. 6.107), the blow pin is inserted through the neck before or after mold closing. The blow pin body has the secondary function of calibrating the bore of the bottle neck. If the blow pin is plunged into the neck after mold closing, it is possible to produce a flash-free bottle mouth.

After the mold opens, the finished bottle is stripped off by retracting the blow pin through a stripper plate. The blowing pin contains channels for the circulation of blowing air and cooling water. There are blow-molded products, such as toys, decorations, and medical components, that do not have internal apertures that can accept blow pins. In such cases, blowing air can be supplied through a blow needle, such as a hypodermic medical needle, that punctures the parison wall after the mold closes (Fig. 6.108).

The needle can be advanced and retracted by means of an air or hydraulic cylinder and is located at a point close to a pinch-off edge. The parison is gripped by the mold at this point and so is unable to deflect away from the advancing needle point. The blow needle leaves an extremely small witness mark that may not be noticeable, or if visible, is often acceptable within the function of the finished part. If not, it can be sealed by a secondary operation.

VENTING AND SURFACE FINISH

When a parison is blown, a large volume of air must be displaced from the mold cavity in a short time. Because blowing is carried out at relatively low pressure, it is essential to provide venting

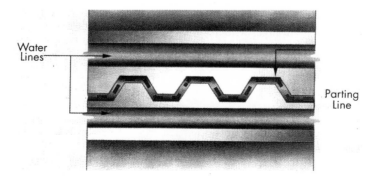

Figure 6.106 Example of a trapezoidal cross-section insert at the parting line

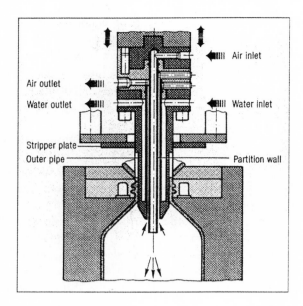

Figure 6.107 Example of a calibrating blow pin

to allow this air to escape without resistance. Unless a gloss finish is required on the molding, it is common practice to sandblast the cavity to a fine matte finish. This helps air to escape as the expanding parison touches the cavity face, but it is not sufficient in itself. Vent slots may be cut at appropriate points into the mold parting face to a depth of 0.05 to 0.15 mm. The appropriate point is where there is a possibility for air to collect as the hot plastic expands in the cavity. Figure 6.109 shows a schematic of an injection molding of a preform mold including venting slots. This type of mold is used after preforms are molded, usually in a multicavity mold as shown in Figure 6.110. Multicavity preform molds for PET can have up to 144 cavities.

Venting can also be provided within the mold cavity by means of inserts provided with vent slots, porous sintered plugs, or by holes with diameters not greater than 0.2 mm (chapter 17). Such holes are machined only to shallow depths and are relieved by much larger bores machined from the back of the mold.

Cooling

Efficient mold cooling is essential for economical blow molding. As in injection molding, typically up to 80% of a blow-molding cycle is devoted to cooling. Molds are constructed as far as possible from high thermal conductivity aluminum alloys, and water-cooling channels are placed as close as possible to the surfaces of cavities and pinch-off zones. Because blow molding is a relatively low-pressure process, the channels can be quite close to the surface and quite closely spaced before mold strength is compromised. The actual dimensions will depend on the heat-transfer rates and cooling temperature requirements for the materials of construction and the plastics being processed (Tables 6.13 to 6.16). As a guide, channels may approach within 10 mm of the cavity and center spacing should not be less than twice the channel diameter.

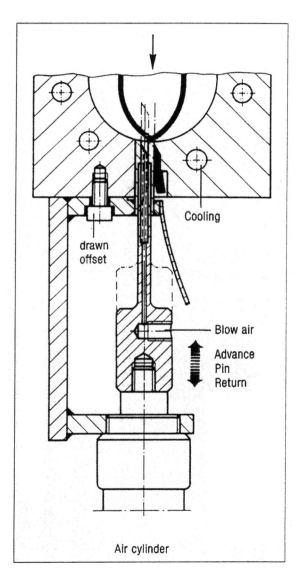

Figure 6.108 Example of blow needle

If the mold body is cast, the cooling channels can be fabricated in copper pipe to closely follow the cavity contours before being cast in place. If the mold is machined, drilling and milling will produce channels and it is not usually possible to follow the cavity contours so closely.

An alternative in cast molds is a large flood chamber (Fig. 6.111). However, efficient water cooling requires turbulent flow, and this may not be attained in a flood chamber or in large coolant channels (chapter 17). Many small channels are better than a few large ones. The cooling circuits will normally be zoned so that different areas of the mold can be independently controlled. The coolant flow rate should be sufficient to ensure turbulent flow and to keep the temperature differential between inlet and outlet to about 3°C.

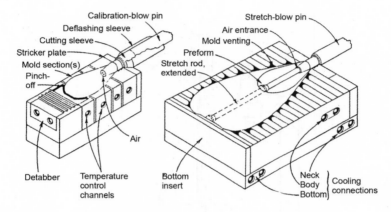

Figure 6.109 Example of air vent slots in an injection molding of a preform mold

Figure 6.110 View of a multicavity preform mold in the background with blow molds and molded bottles in front (courtesy of SIG Plastics International)

Process or System	Heat-Transfer Coefficient (Btu/°F/h/sq ft)	Limiting Factors
Injection molding	30	Thickness and area available
Blow molding	25	One-side cooling and pinch-off cooling
Foam injection molding	20	Thickness and cellular structure
Jacketed vessels	50	Agitation and area available
Bath or trough	100	One-side cooling, thickness, and area available
Shell and tube exchanger		
Oil to water	100	Area and temperature difference
Water to water	250	Area and temperature difference

Table 6.13 Cooling characteristics

Molds on blow molding machines	20–55°F
Molds on bottle molding machines	50–55°F
Eddy current drive	80–85°F
Extruder barrels	80–85°F
Hydraulic oil coolers	80–85°F
Air compressors	80–85°F
Vacuum pumps	80–85°F
Refrigeration condensers	80–85°F
Temperature-control units	80–85°F
Molds on injection molding machines	20–55°F
Extruder troughs and cooling baths	50–55°F

20°F (−7°C); 50°F (10°C); 55°F (13°C); 80°F (27°C); 85°F (29°C); 250°F (121°C).

Note: High-temperature cooling requirements for special compounds may require temperatures in excess of 250°F.

Table 6.14 Cooling temperature requirements

Plastic	Recommended temperature	
	(°C)	(°F)
Polyacetates	80–100	176–212
Polyamides	20–40	68–104
Polyethylenes and PVC's	15–30	59–86
Polycarbonates	50–70	122–158
Polymethyl methacrylates	40–60	104–140
Polypropylenes	30–60	86–140
Polystyrenes	40–65	104–149

Table 6.15 Examples of blow-molding mold cavity temperatures based on plastic being processed

Cooling Analysis	
Mold surface Temperature	Part quality: even cooling prevents distortion
Freeze time	Minimizes cycle time.
Coolant temperature and flow rate	Optimizes coolant: can eliminate need to chill coolant.
Metal temperature	Cooling efficiency: ensures optimum circuit design.
Warpage Analysis	
Warped shape	Tendency to warp.
Single variant Warpage shape	Indicates fundamental causes of warpage.
Flow Analysis	
Fill pattern	Weld line position.
	Air trap position.
	Position of vents.
	Overpacking: excessive costly part weight.
	Overpacking: warpage due to differential shrinkage.
	Underflow: structural weaknesses.
Pressure distribution	Pressure required to fill.
	Clamp force required.
	Overpacking: ribs, etc. sticking in mold.
Temperature	Poor surface finish.
	Weak weld lines.
	Distortion due to differential cooling.
Shear stress distribution	Quality of part: tendency to distort.
	Quality of part: tendency to crack.
	Cycle time: low stresses permit hotter demold temperature.
Shear rate	Avoids degradation of material.
cooling time	Shows tendency to distort due to uneven cooling.
Flow-angle	Quality of part: molecular orientation.
packing pressure	Under/overpack to poor packing.
volumetric shrinkage	Dimensional variations due to poor packing.

Table 6.16 Examples of computer software information generated and typical problems it can solve (chapter 25)

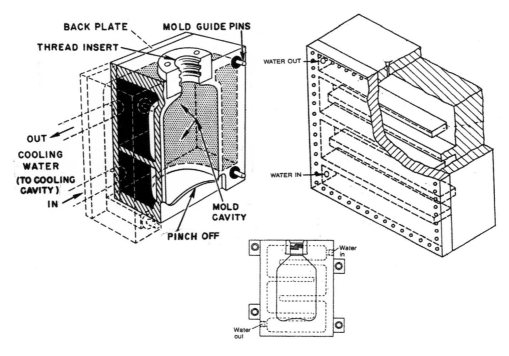

Figure 6.111 Examples of water flood cooling blow-molding molds

PLASTIC MATERIAL

Many thermoplastics can be blow molded. Different properties can be obtained with the different available and useful plastics that are now used to blow mold different sized and shaped products, providing the machines are designed to process the plastics (Tables 6.17 and 6.18). About 50 wt% is HDPE, 25% PET, 20% PVC, 2% polypropylene, and 6% other plastics are blow molded.

Properties can include formability, weight, strength, toughness, clarity, gas barriers, chemical resistance, heat resistance, color, and many others. Examples of these properties are in Tables 6.19 to 6.26 and Figures 6.112 and 6.113. Different combinations of properties can be achieved by combining two or more plastics. Parisons and preforms can be coextruded, coated, dipped, coinjected, and so on in order to provide the combinations of different properties available in different plastics. Stretch blow molding also provides an important processing technique to improve properties and reduce the quantity of plastics to improve cost/performance characteristics.

Examples of properties that can be obtained with the different plastics are as follows: clarity (crystal PS, AN, styrene AN, PVC, stretched PET, PSF), moisture barrier (PP, HDPE, transparent nylon), oxygen barrier (PET, PVOH, PVDC, PVC, AN, PSF, transparent nylon), and so on.

Table 6.17 Examples of properties of thermoplastic bottles

Properties	Acrylic multipolymer	Nitrile	Polycarbonate	Polyester (oriented) (PET)[b]	PETG copolyester	Polyethylene Low density	Polyethylene High density
Resin density	1.09–1.14	1.15	1.2	1.35–1.40	1.27	0.91 to 0.925	0.94 to 0.965
Clarity[a]	clear	clear	clear	clear	clear	hazy transparent	hazy translucent
Permeability to water vapor	high	moderate	high	moderate	moderate	low	very low
Oxygen	low	very low	moderate to high	low	low	very high	high
CO_2	moderate	very low	moderate to high	low	low	very high	high
Resistance to acids	poor to good	poor to good	fair	fair to good	fair	fair to very good	fair to very good
Alcohol	good	fair	fair	good	good	good	good
Alkalis	poor to fair	good	poor to fair	good	poor to fair	good to very good	good to very good
Mineral oil	good	very good	good	good	poor	fair	good
Solvents	poor	good	poor to fair	good	poor to good	poor	fair
Heat	fair	poor to fair	very good	poor to fair	fair	poor to fair	fair to good
Cold	poor	fair	good	good	good	very good	very good
Sunlight	good	fair	good	good	fair	fair	fair
Temperature at which finished product distorts[d]	180°F to 195°F	140°F to 150°F	260°F to 280°F	100°F to 160°F	160°F to 220°F	160°F to 220°F	160°F to 250°F
Stiffness	moderate to high	moderate to high	high	moderate to high	low	low	moderate
Resistance to impact	poor to good	high	excellent	good to excellent	excellent	good to very good	good to very good
Unit cost	moderate to high	high	very high	moderate	moderate to high	low	low
Typical uses	foods, drugs, cosmetics	cosmetics, household chemicals	cosmetics, foods, water, milk	carbonated beverages, mouthwash, liquor, edible oil, drugs, cosmetics	cosmetics, foods, mineral oil, personal care	cosmetics, personal products, mustard, drugs	detergents, bleaches, milk, chocolate, syrup, industrial cleansing powders, drugs, cosmetics, lube oil, edible oil

[a] Bottles made from any of these resins are available in opaque colors
[b] Polyethylene terephthalate
[c] Polymethylpentene
[d] Range given here is limiting temperature in normal use

Table 6.17 (continued)

Properties	Polypropylene Regular	Polypropylene Oriented	Polystyrene	Polysulfone	SAN (styrene acrylonitrile)	TPX[c]	PVC
Resin density	0.89 to 0.91	0.90	1.0 to 1.1	1.24	1.07 to 1.08	0.83	1.35
Clarity[a]	moderate haze	clear	clear	clear	clear	moderate haze	clear
Permeability to water vapor	very low	very low	high	high	high	very low	moderate
Oxygen	high	high	high	low to moderate	high	very high	low
CO_2	moderate to high	moderate to high	high	low to moderate	high	very high	moderate
Resistance to acids	fair to very good	fair to very good	fair to good	fair to very good	high	high	low
Alcohol	good	good	fair	fair to good	poor	fair to good	good to very good
Alkalis	very good	very good	good	very good	good	very good	good to very good
Mineral oil	fair	fair	very good	very good	fair	fair to good	good to very good
Solvents	poor to good	poor to good	poor	fair to good	poor	poor to good	good
Heat	good	good	fair	very good	fair	very good	poor to good[e]
Cold	poor to fair	very good	poor	very good	fair	fair	poor to fair
Sunlight	fair to good	fair to good	fair to poor	fair	fair to poor	poor to fair	fair
Temperature at which finished product distorts[d]	250°F to 260°F	250°F to 260°F	200°F to 220°F	330°F to 365°F	170°F to 190°F	230°F to 260°F	140°F to 150°F
Unit cost	moderate	moderate to high	moderate	very high	moderate to high	high	moderate to high
Resistance to impact	poor to good	very good	poor to good	good to very good	poor to good	poor to good	fair to good
Stiffness	moderate to high	moderate to high	moderate to high	high	moderate to high	moderate to high	moderate to high
Typical uses	drugs, cosmetics, syrups, juices, detergents, mouthwash, shampoo, IV solutions	detergents, drugs, mouthwash, shampoo, household chemicals, IV solutions, liquid soaps	dry drugs, petroleum, jellies, vitamins, spices	laboratory ware, microwave cooking	dry drugs	laboratory ware, microwave cooking	cosmetics, personal care, household chemicals, edible oils, vinegar

[a] Bottles made from any of these resins are available in opaque colors
[b] Polyethylene terephthalate
[c] Polymethylpentane
[d] Range given here is limiting temperature in normal use
[e] Ketones; esters, chlorinated and aromatic solvents: poor; aliphatic solvents: good

	Resin				
Criterion	PET	PET copolymer	Lowmonomer PVC	Nitriles	Poly-carbonate
Taste					
Brown goods	A	A	UA	M	M
White goods	A	A	UA	–	–
Odor	A	A	UA	M	M
Appearance: clarity, color, transparency	E	A	A	M	A
Strength	A	A	A	–	A
Ability to produce handleware	No	No	Yes	Yes	Yes
Product retention					
Alcohol loss	A	Note[b]	A	A	Note[b]
Proof rise (water loss)	A	M	Note[c]	UA	UA
Approval by					
FDA	Yes	Note[b]	Yes	–	Yes
BATF	Yes	No	No	No	No
Resin cost, cents per pound (December 1982)	60	–	70	92 to 108[d]	169[d]

[a] E = excellent; A = acceptable; UA = unacceptable; M = marginal
[b] Unknown, [c] Slightly higher, [d] Excessive

Table 6.18 Examples of various plastics suitable for plastic liquor bottles

Properties	Desired goals
Stiffness	Varies with product requirements
Environmental stress crack resistance	Generally maximum, though sometimes not important
Gloss	Usually high, but varies with product requirements
Overall appearance	High gloss, smooth surface, no flow lines
Wall thickness uniformity	Minimum gauge variation in walls
Weld-line thickness	Equal to or greater than the side wall thickness
Parting-line difference	Minimum gauge variation from side wall
Mold cycle time	Minimum

Table 6.19 Important properties of extrusion blow-molded products and the desired goal(s) for each

Extrusion blow molding represents the biggest outlet for blown containers, with HDPEs being the major plastics used. Recognize that there are different grades of HDPE that include those with changes in density, stiffness, strength, hardness, toughness, stress crack resistance, permeability, creep resistance, softening temperature, and gloss.

	Higher Density	Higher Melt Index
Bottle[1] Properties		
Stiffness	Large Increase	NSE*
Environmental Stress Crack Resistance	Decrease	Large Decrease
Gloss	Large Increase	Increase @ 0.915 g/cm³; NSE @ 0.932 g/cm³
Overall Appearance[2]	Improvement	NSE
Wall Thickness Uniformity	NSE	NSE
Weld Line Thickness	Large Decrease	NSE
Parting Line Difference	NSE	Large Increase at high blow pressure
Mold Cycle Time	Large Decrease	NSE
Test limited to range specified:	Densities of 0.915 to 0.932 g/cm³	Melt Indexes of 1.3 to 4.7 g/10 min.

*NSE = no significant effect
[1] tests made on 4 oz. (115 ml) Boston round bottles
[2] quantitative rating of gloss, surface smoothness and flow lines in bottles

Table 6.20 Changes in extrusion blow-molded bottle properties resulting from resin properties

Table 6.21 Changes in extrusion bold-molded blow properties resulting from changes in extrusion and molding conditions

Bottle[1] Properties	Extrusion Conditions			Molding Conditions		
	Higher Melt Temperature	Higher Annular Die Clearance	Higher Extrusion Rate	Higher Mold Temperature	Higher Molding Time	Higher Blow Pressure
Stiffness	NSE*		Slight Decrease	NSE	Increase	NSE
Gloss	Increase	NSE	NSE	Improvement	NSE	NSE
Overall Appearance[2]	Slight Improvement	NSE	NSE	Improvement	NSE	Improvement
Wall Thickness Uniformity	Slight Decrease	Decrease	NSE	Slight Decrease	NSE	NSE
Weld Line Thickness	NSE	Large Increase	NSE	NSE	NSE	Large Increase from 10 to 50 lb/in.[2]
Parting Line Difference	Slight Increase	NSE	NSE	Increase	NSE	Large Increase
Mold Cycle Time	Increase	Increase	NSE	Large Increase	Large Increase	NSE
Tests limited to ranges specified	Temperatures of 150-170°C (300-340°F)	54.5 to 79.5 mils (1.4 to 2.0 mm)	0.2 to 0.4 in./sec. (5 to 10 mm/sec)	15 to 70°C (60 to 150°F)	12 to 24 sec	10 to 80 lb/in.[2] (0.7 to 6.5 kg/cm[2])

*NSE = no significant effect
[1] tests made on 4 oz. (115g) Boston round bottles
[2] quantitive rating of gloss, surface smoothness and flow lines in bottles

Type of bottle	Rate ($m^2\,day^{-1}$)	
	Oxygen (cm^3)	Water vapor (g)
PET (oriented)	10.2	1.10
Extrusion blow molded PVC	16.4	2.01
Stretch blow molded PVC (impact-modified)	11.9	1.8
Stretch blow molded PVC (nonimpact-modified)	8.8	1.3

At 38 °C (100 °F).

Table 6.22 Gas barrier transmission comparisons for a 24 fl oz (689 cm^3) container weighing 40 g

Type of bottle	Percent
Extrusion blow molded PVC	–
Impact-modified PVC (high orientation)	4.2
Impact-modified PVC (medium orientation)	2.4
Impact-modified PVC (low orientation)	1.6
Nonimpact-modified PVC (high orientation)	1.9
Nonimpact-modified PVC (medium orientation)	1.2
Nonimpact-modified PVC (low orientation)	0.9
PET	1.2

Seven days at 80 °F (27 °C).

Table 6.23 Volume shrinkage of stretch blow-molded bottles

Position	Direction	Yield point (kg/mm^2)	Ave.	Break point (kg/mm^2)	Ave.	Elongation at break (%)	Ave.
Upper panel	Axial	9.30 / 9.47	9.30	8.82 / 9.72	9.27	55.0 / 57.5	56.3
	Hoop	– / –	–	18.29 / 19.37	18.83	22.5 / 25.0	23.8
Lower panel	Axial	9.33 / 9.28	9.31	11.86 / 8.62	10.24	92.5 / 32.5	62.5
	Hoop	– / –	–	18.58 / 19.66	19.12	20.0 / 25.0	22.5

Table 6.24 Tensile test data of PET plastic

Beverage bottles provide the major reason for PET being the second major plastic consumed in stretched injection blow molding. PET properties meet performance requirements for pressurized beverage bottles.

PVC is a major plastic used in great quantity in the United States for extrusion blow molding due to its low oxygen transmission, water-clear clarity, relative ease of processing, low cost, and so on. It is being challenged by stretch-blown PP and stretch-blown PET. PP and PP copolymers are used in packaging containers for cosmetics, pharmaceuticals, and some foods (syrup, etc.), especially for medical-surgical fluids, where hot filling is required, and so on.

This brief review on the different plastics is used to emphasize the importance of using the correct plastic to meet specific requirements. The real test on any material is when it is blown, either unoriented or stretched oriented. Details on properties and melt processabilities are reviewed in chapters 1 to 3.

Blow Molding and Plastic

Different plastics can require different blow molding equipment, such as feed mechanisms, screw designs, and mold designs (chapters 3 and 17). Here is an example concerning polypropylene (PP). Blow molding processes only about 2 wt% of PP. The combination of economy and good properties offered by PP leads to its widespread use in films, fibers, and injection moldings. So, why are the packaging advantages of PP films not exploited in the containers and bottles that the blow-molding process produces? Why has the outstanding environmental stress crack performance of PP, along with its chemical resistance, not made it the material of choice for blow molding?

The answer lies in the development of blow-molding machinery and the differences in melt rheology between PP and the popularly used PE. Blow molding's rapid developments in machinery

Plastic	Temperature (°C)
LDPE	130–180
MDPE	150–200
HDPE	160–220
HMWPE	180–230
PVC	190–205
PP	200–220
PS	280–300
PA	240–270
POM	150–280
ABS	180–230
ABS/PC	230–250
PPE	240–250
PBT	245–260
PBT/PC	240–260
PUR	180–190

Table 6.25 Guide to plastics processing temperatures for blow molding

and technology began in the late 1930s, virtually contemporary with the introduction of PE. PP was not commercialized until the mid-1950s, so BM technology was largely optimized on the requirements for PE (chapter 29).

Blow molding is a relatively low-pressure process, using only 1% or less of the pressure levels employed in injection molding. This requires a low melt viscosity, produced by high shear rates and temperatures. The difficulty with PP is that its melt viscosity is far more sensitive to temperature and shear rate than is the case with PE, so there are problems in working PP on PE equipment (221).

To change this condition, machinery and technology are now being developed specifically for PP. At the same time, developments in PP are making it a more suitable material for the process. The increased transparency of nucleated, clarified, and random copolymer grades is making them more attractive for bottle and container applications. Advances in polymerization technology have

Increase in these factors affects these properties	Resin Properties		Extrusion Conditions		
	Density	Melt Index	Melt temperature	Annular die clearance	Extrusion rate
Stiffness	Large increase	—	—	—	Slight decrease
Environmental stress crack resistance	Decrease	Large decrease	—	—	—
Gloss	Large increase	Increase at 0.915 density / No effect at 0.932 density	Increase	—	—
Overall appearance	Improvement	—	Slight improvement	—	—
Wall thickness uniformity	—	—	Slight decrease	Decrease	—
Weld-line thickness	Large decrease	—	—	Large increase	—
Parting-line difference	—	No effect at low blow pressure / Large increase at high blow pressure	Slight increase	—	—

Table 6.26 Examples of fabricating conditions on blow-molded PE bottles

made it possible to produce grades with melt strengths and shear dependencies that are more tolerant of the blow-molding process. Consequently, blow molding PP is a growth area. Its market share is currently small, but its growth rate is second only to PET. PP blow molding is targeted to take market share from PVC and high-density polyethylene. The continuous extrusion blow-molding process is the most widely used for working with PP.

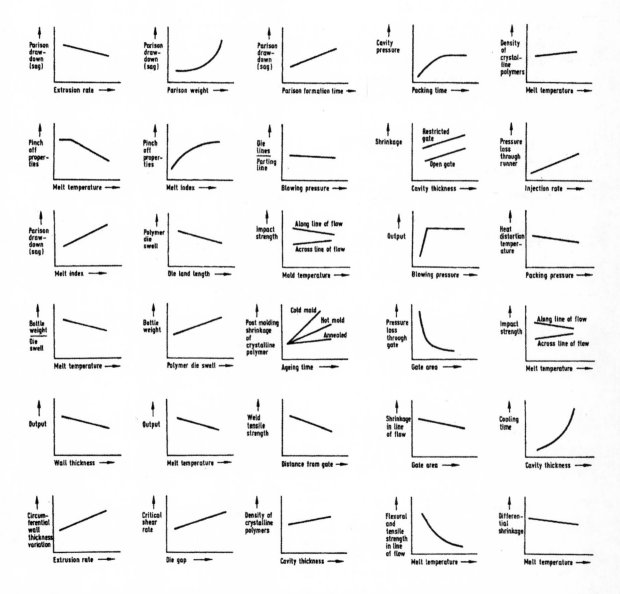

Figure 6.112 Examples of effects of the blow-molding extruder and plastic variables on product performances

Stretch blow-molded PP containers have some important advantages. They can be sterilized at higher temperatures than competitive containers and are not prone to environmental stress cracking. Stretched PP containers can be hot filled at about 90°C.

BEHAVIOR OF PLASTICS

A representative cross-section of the myriad of plastics available to designers has been described in this chapter and throughout this book. This section is to acquaint the designer with the structural behaviors of plastics. It provides concepts as background for estimating and anticipating behaviors in design situations. Obviously, there are no universal methods to describe the behavior of all plastics, just as there is no single reference that completely covers the behavior of all metals and their alloys. Nonetheless, all plastics show many behavioral similarities. The differences are frequently in terms of magnitude, not of kinds.

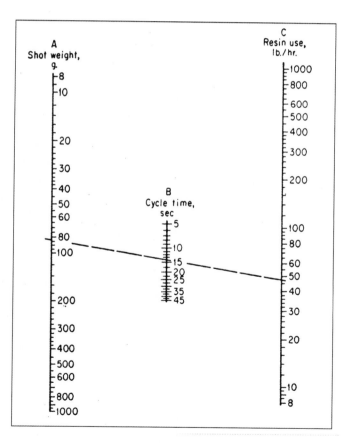

Figure 6.113 Nomogram for injection blow-molded preform shot weight, cycle time, and resin use

The stress strain and strength behavior of plastics vary widely, depending on the generic type or family of plastic and on the specific composition of compounds within that family. Many factors interact to alter behavior of a given compound over a very wide range. The following are key parameters that must be considered in structural design of plastics:

1. *Process*: The blow-mold process used to convert plastic materials into structures may dictate the structural performance of the finished product, based on the following:

    a. Orientation of the molecular structure of a thermoplastic may strengthen the product in the direction of orientation. Such preferential orientation can be used to advantage; however, the designer must be on guard to ensure that orientation does not prove to be the cause of failure. For example, flow patterns and knit and weld lines developed during molding can cause orientation that creates zones of weakness.
    b. Oxidation and crystallization induced during processing can embrittle an otherwise ductile material. Polyethylene, for example, held too long at high temperatures may either oxidize in the air or develop excessive crystallization.

2. *Magnitude and duration of stress, strain, and temperature*: At a given temperature, both the magnitude and duration of stress or strain affects structural response and strength behavior. Conversely, at a given magnitude and duration of stress or strain, a shift in temperature can produce marked changes in structural response and strength behavior.

3. *Environment*: The environment interacts with the magnitude and duration of stress, strain, and temperature to further alter material response and strength of plastics. Chemical environment permeability, ultraviolet (UV) radiation, sustained elevated temperature, and even water, for example, can have profound influence on performance and, hence, may dominate the design problem. Environmental effects, or the failure to properly design for them as they interact with sustained stress or strain, have been chief causes of failure of plastics products.

4. *Additives and modifiers*: Fillers and plasticizers alter the basic response and strength of plastic materials. Particulate fillers (wood flour, flake, clay, limestone, etc.) are introduced to reduce resin shrinkage, increase stiffness, improve processing characteristics, or to lower cost. Most fillers also lower impact resistance. Plasticizers increase flexibility and toughness; impact modifiers, such as rubber blends, are more permanent and improve toughness without significant sacrifice in stiffness. Thus, the properties of a basic generic plastic, such as PVC, can be varied to provide either rigid sewer pipe or flexible rubberlike water stops and liners.

5. *Reinforcement*: Stiff strong fibers and flakes incorporated into plastics improve stiffness, strength, and dimensional stability. An increase in their proportion relative to the resin matrix results in a corresponding improvement in these properties.

6. *Similarities with conventional materials*: The plastics behavior introduced here is not necessarily any more complicated than the behavior of conventional structural materials. Overall, the engineer must design both plastics and conventional materials to meet the criteria of time, temperature, and environment. The similarities are as follows:

   a. Time: In designing with conventional materials (steel, aluminum, glass, wood, and concrete), the effects of load duration are recognized in terms of creep and ultimate strength. Time effects are frequently more prominent in plastics, particularly in terms of strength.
   b. Temperature: In designing with steel and glass, brittleness at low temperatures and loss of yield strength at high temperatures must be considered. Similarly, in designing with plastics, both stress strain and strength behavior can vary with temperature, and the low and high temperature limits of certain plastics are frequently in the range of the normal outdoor environmental limits.
   c. Environment: The strength, stiffness, dimensional stability, and useful life of wood are strongly impacted by moisture content. Steel rusts, glass cracks, aluminum corrodes, and so on. The problem is similar for plastics, but because composition can vary widely, each plastic compound must be considered for its strengths and weaknesses on exposure to various environments.

While there are certain similarities, there are significant differences between designing with plastics and with conventional materials (steel, wood, etc.). Stringent specifications define the characteristics, and formal detailed rules define design procedures, for a limited number of aluminums, steels, and glass as well as a broad range of timber species. Concrete and steel quality is checked by fairly simple tests; wood quality is assessed visually in accordance with long established "grading" rules. Working rules of thumb, which may be highly empirical, have evolved from both experience in the field and research using these materials of construction (chapter 19).

Design criteria for plastics are established in different standards and specifications (chapter 22). A myriad of different materials and variations are available (chapter 2).

BARRIER PLASTIC

As briefly reviewed ("*Coextrusion and Coinjection*"), there are barrier requirements for blown products that range from packaging foodstuff to gasoline. The barrier requirements to meet package needs vary depending on the product being packaged. Properties that may need to be retained or excluded can include gases, water vapors, aromas, tastes, and solvents. The result has been the development and use of a wide diversity of barrier plastics in blow molding. Information on coextruding or coinjecting these barrier materials is reviewed in chapters 3 to 5.

Many oxygen-sensitive food products can be either hot filled or retorted. In these cases, the container must have requisite thermal as well as barrier properties. Further, if the container is to be subjected to elevated temperatures during filling or processing, it must be designed to either resist or anticipate paneling. Both of these factors can considerably increase the complexity of developing satisfactory rigid barrier containers.

An example is the success of packaging carbonated beverages using PET plastic. The plastic bottle industry continues its research into expanding plastic bottles in other large-volume applications to replace glass and metal with blow-molded plastic containers. There are more high-volume markets that could be expanded and/or switched into plastic containers, such as beer (United States).

To date, it appears that Americans prefer their beer in glass bottles, but that could change if various development programs take off, such as PINTA. As reported, when Rob Peabody had a few friends over to watch a football game on TV, he served them beer in plastic bottles. As head of the newly formed Polyester Intermediates Americas (PINTA), Peabody figures on plastic beverage bottles to be a strong growth market in North America (275). The BP PINTA's business unit is to produce purified terephthalic acid (PTA), a raw material used to make polyethylene terephthalate (PET), a primary ingredient in plastic bottle manufacturing.

Although plastic bottles do not yet dominate U.S. beer sales, they reign supreme as holders of everything from juice to water. In fact, 60% of any plastic beverage bottled in North America is made with BP PTA. The driving force for this application is to keep bottle contents fresh longer. That includes BP's dimethyl 2,6-naphthalenedicarboxylate (NDC), which provides heat resistance for pasteurizable plastic beer bottles, and Amosorb, which keeps oxygen out of sensitive beverages, such as juice. Both are produced by PINTA in Naperville, Illinois.

There are at least four basic ways in which to increase the barrier properties of a rigid plastic container. The first is to improve the barrier properties of the plastic being used in a monolayer container. The second is to place a barrier coating on either the inside or the outside of the monoplastic layer. The third is to employ several distinct plastics in a multilayer coextruded structure with each providing a different specific barrier or performance quality.

An example of coating is from Tetra Pak USA in Vernon Hills, Illinois, which has FDA acceptance for its Glaskin silicon dioxide barrier coating technology for PET containers (542). Special plasma technology deposits a clear, glasslike coating, thinner than a human hair, on the bottle interior. Studies show that the coating increases the oxygen barrier of a PET bottle seventeenfold and the CO_2 barrier twenty-five-fold. The coating can extend the shelf lives and maintain the taste and nutritional values of beverages, such as beer, carbonated soft drinks, and fruit juices. The Glaskin coating has a shelf life of more than a year. The very thin coating offers high elasticity and perfect clarity with no haziness. Since it is inside the bottle, it cannot be damaged by conveyors and packaging lines. Initial use started in Europe, where Spendrups and Bitburger beer are bottled in PET with the Glaskin coating.

The following plastics are being stretched coinjected blow molded with PET plastic for developing beer applications: EVOH (ANC), nylon MXD6 (Schmalbach Lubeca), nylon MXD6 (Claropak), oxygen scavenger (Continental), and Amosorb 02 Scavenger (Kortec).

Monolayer Chemical reactive systems are used to change the monolayer barrier behavior of certain plastics. As an example, fluorination, a secondary process used on monolayer HDPE containers, involves the additional step of treating the container with fluorine gas. This treatment adds side chains of fluorine to the molecular structure of the HDPE, resulting in improved barrier properties.

Examples of fluorination processes used include Air Products & Chemicals' process that injects fluorine gas during the blowing cycle. The Union Carbide/Linde process is a postmolding treatment in which molded bottles are placed in a chamber and exposed to fluorine gas at elevated temperatures. Both processes are accepted by the industry, and fluorinated HDPE bottles are being used for packaging pesticides, paint thinners, charcoal lighter fluid, kerosene, and so on.

There is also Linde's Surface Modified Plastics (SMP) process. It was a spin-off from their aerosol container research. The process is unique among current fluorination methods for polyethylene because it allows precise control over the degree of treatment. In effect, the user is able to custom build a barrier that is specifically suited for the application at hand.

The development of barrier plastics that can be used in monolayer containers to pack oxygen-sensitive food products has not, to date, been commercially successful. As an example, plastic producers have developed barrier PET copolymers, but in all cases, they have had to trade off some other critical performance characteristics, such as impact strength or creep resistance. Further, barrier plastics, even when they do achieve acceptable balances of characteristics, are too expensive, running at least between $1 and $3 a pound. Very little development work is being done in this area and the likelihood of an inexpensive barrier PET is several years off at best.

Coating The concept of coating a plastic to improve its barrier performance is not new. Some of the first and still most widely used flexible barrier packages are PVDC-coated polyester and oriented polypropylene. Coating is perceived by many as the most reasonable low-cost method to improve the barrier properties of rigid monolayer containers.

Coating has several drawbacks. The primary functional problem is adhesion. Another is the cost of high-speed continuous coating lines. Also, there is the problem of achieving an adequate barrier with a single pass. Assuming one is coating the outside of a container, there are three basic types of coating application systems: roll coating, dip coating (with or without a wipe step), and spray coating.

Roll and dip coating are used in Europe to apply a layer of PVDC directly on oriented PET (OPET) bottles for beer. In the United States, these application methods have been used to coat both preforms and finished PET bottles. Coating the preform appeared originally to have some advantages because the subsequent stretch-blowing step oriented the PVDC and increased its barrier properties. During 1980, wine bottles used coated preform. There were, however, resultant adhesion problems and an overthinning of the PVDC layer during blowing, which effectively reduced the net barrier effectiveness despite the biaxial orientation achieved. In addition, the lines were too slow to be effective commercial processes.

There is also the approach to achieve an adequate barrier with a single pass. Efforts have been made to develop dual-pass systems or coating systems in which a superior oxygen barrier (EVOH or EVA) is covered with a water barrier, but the cost has outweighed the benefits. The critical point to make is that although coating does not produce a totally successful barrier container, it is a commercially available technology that offers a modest degree of barrier improvement.

Some of the other emerging coating technologies that may generate interest in the future are other sprayable barrier materials, including EVOH, and sinter coating techniques that would permit the use of certain plastics that can be powdered but not sprayed.

Multilayer Most of the development efforts to produce high-quality rigid barrier containers involve the use of multilayer structures. There are two basic approaches to the production of multilayer structures. These are coextrusion and coinjection. The basic issues to consider when developing a multilayer barrier container are which barrier material to use and which manufacturing technology to employ. Blow molding multilayer hollow products continues to experience a phase of worldwide interest, including a rapid upsurge. It utilizes the material-specific properties of various plastics for the production of high-grade multilayer containers.

The introduction of the five-layer to seven-layer technique in the United States brought the breakthrough in 1984 for commercial production of these containers, particularly in the foods industry (ketchup, mayonnaise, soup concentrates, baby food, and the like). A crucial factor was the development of a completely new multilayer head extrusion technology permitting wall thicknesses down to less than 0.03 mm (0.001 in) with excellent, constant coaxial wall thickness distribution. This was the key to using expensive barrier and bonding agent materials at commercially viable production costs.

The advantages of these bottles are many. Here are some of them, related to different material combinations: (a) excellent barrier properties against gases, such as nitrogen, carbon dioxide, and air, as well as against water vapor (foodstuffs, pharmaceuticals), (b) resistance to aroma transmissions (cosmetics), (c) protection against UV radiation (mild), (d) higher strength (greater product stackability), (e) lightweight (lower transport costs), (f) scratch-resistant glossy surfaces (cosmetics), (g) resistance to aggressive media (insecticides, herbicides), (h) barrier to solvents (paint industry), (i) easy printability and clear printed impressions (cosmetics), and (j) prevention of static charging (explosion protection).

Figure 6.114 is an example that shows the relation between comonomer concentration, the moisture transmission, and gas transmission properties of these copolymers. These exceptional barrier properties allow the use of extremely thin layers of these materials in combination with other inexpensive plastics to BM economical high-performance products. Table 6.27 indicates the range of commercially available EVOH plastics. Within this range, conventional melt BM processes readily process EVOH plastics.

The method that has the greatest application potential is to arrange the barrier layer in the middle between the inner and outer body layers. Whether reclaimed material is used in the outer layer is a quality question that the container filler has to answer based on product performance requirements. The seven-layer combination with reclaim in the penultimate layer from the outside, without any further bonding agent to the body material, is the most satisfactory solution (economically) to date.

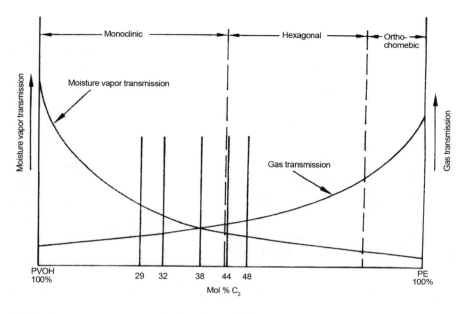

Figure 6.114 Comonomer concentrations vs. barrier properties of crystalline structures

Melt index (g/10 min)	Density (g/cc)	Ethylene content (mol %)	Melting point (°C)
0.7–20	1.13–1.21	29–48	158–189

Table 6.27 EVOH plastic range of properties

With the coextruded technology well established and the hardware in position, the variety of achievable properties is readily extendable through the correct combination of different materials. The potential for BM packaging products to be much more than simple bottles has been known for some years. Now the know-how and economics are such that many previously futuristic ideas are much closer to reality.

The potential in coextruded bottles, coupled with the machinery platform being established, means that glass and metal will come under increasing pressure. Often, the higher initial costs of the plastic materials are counterbalanced by the total processing and packaging economics that can be obtained. The impetus behind this rate of growth is the combination of the inherent versatility of plastics with blown shapes (cylindrical to noncylindrical, irregular hollow-shaped industrial products, etc.) that provides cost advantages over other processes.

The advantages of coextrusion are that the equipment and technology are readily available, relatively high speed, and economical. The disadvantages arise from the subsequent converting process in which significant amounts of scrap are generated, wall and flange dimensions tend to be difficult to maintain, and there could be an exposed cut edge that may be critical in situations where an improperly molded water-sensitive barrier material (EVOH) is exposed to long periods of high relative humidity. Various attempts have been made to resolve the scrap problem. Coextrusion offers a greater variety of conversion options than does coinjection at lower processing cost.

BARRIER MATERIAL TYPE

There are four popular basic categories of barrier materials currently being used by producers of multilayer containers. These are EVOH, PVDC, EVA, and nylon. Most attention is focused on the first two of these, with EVOH having the most continual, significant current prospects because of its greater processing window and compatibility. PVDC remains the preferred material for flexible packaging and has the advantage of being unaffected by moisture. PVDC is, however, basically difficult to process and recycle because of its thermal sensitivity. EVA is an excellent barrier but extremely hygroscopic, which severely limits its use (Table 6.28).

To date, there is no dominant technology, product, market, or set of performance specifications directing the development of rigid plastic barrier containers. They deal with a number of variables that include choice of materials, choice of conversion processes, choice of filling and processing technology, and redefinition of product performance requirements. There is no single optimum container. Rather, there is an increasing diversity of rigid barrier plastic containers, each designed for a specific end-use application. This situation does not exist with the OPET (oriented PET) soft drink container business.

BLOW MOLDING REINFORCED PLASTIC

Practically all BM uses unreinforced TPs. In the past, different systems were used to produce RP BM products. One of the latest is from Krupp Kautex, which uses two extruder blow-molding

Polymer	O_2 transmission rate (25°C, 65% RH cc.mil/100 in^2/24 hrs)	Moisture vapor trans. rate (40°C, 90% RH gm.mil/100 in^2/24 hrs)
Ethylene vinyl alcohol	0.05–0.18	1.4–5.4
Nitrile barrier resin	0.80	5.0
High barrier PVDC	0.15	0.1
Oriented PET	2.60	1.2
Oriented nylon	2.10	10.2
Low density polyethylene	420	1.0–1.5
High density polyethylene	150	0.69
Rigid PVC	5–20	0.9–5.1

Table 6.28 Examples of barrier properties of commercially available plastics

(EBM) methods to provide increased structural properties through the use of relatively long glass-fiber reinforcement, such as with PP plastic (chapter 15). As an example, the 15 wt% content uses 7-mm-long (2-in-long) fibers.

One approach extrudes unreinforced melt into a low-shear Z-blade or sigma-blade batch kneader. Dry, chopped glass fibers are blended into the melt under vacuum, and a discharge screw in the mixer's bottom feeds the compound to the blow molder's accumulator head. Another approach uses commercially available glass-fiber/PP-compounded pellets made by the pultrusion process. Pellets are plasticized using a special screw that melts by heat conduction rather than by shear in order to minimize fiber damage. The parison is blown initially at very low air pressure; after complete expansion, full pressure is used.

DESIGN

Designers should become aware of the potential that blow molding offers, since shapes ranging from simple to intricate can be fabricated. The process is especially amenable to the designer's goal of consolidating as much function as possible into a single product. Some of the features that can be incorporated include threads, inserts, fasteners, snap fits, hinges, and others somewhat similar to those covered under injection molding.

Design is the process of devising a product that fulfills as completely as possible the total requirements of the user, and at the same time, satisfies the needs of the fabricator in terms of cost effectiveness (return on investment). The efficient use of the best available material and production process should be the goal of every design effort, including tool design. Product design is as much an art as a science. Design guidelines for blow-molding plastics have existed for over a century, pro-ducing many thousands of parts meeting service requirements, including those requiring long lives.

Designing acceptable products requires knowledge of plastics that includes their advantages and disadvantages (limitations) and some familiarity with processing methods. Until the designer becomes familiar with processing, a qualified fabricator must be taken into the designer's confidence early in development and consulted frequently during those early days. The fabricator and mold or die designer should advise the product designer on the materials' behaviors and how to simplify the product in order to simplify and/or reduce cost of processing. Understanding only one process, and in particular just a certain narrow aspect of it, should not restrict the designer.

The effective exploitation of product design opportunities is the key start to success. In turn, success hinges on other factors, such as the proper selection of materials and using the best available processing equipment. Detailed analysis to the design approach is given in chapter 19.

Because new materials and equipment continue to be more productive and produce better quality products, one should stay abreast of new material and equipment developments and evaluate them logically. With designing, there is an extremely vast area for improving profitability by ensuring that the best available material and best processing method are used to meet specific design performance requirements. Fabricating plastic products is analogous to modern living: change continually occurs.

The aim of perfection requires constant change; however, recognize that perfection is never reached (chapter 30).

As reviewed at the start of this chapter, blow-molded plastic products have certain inherent characteristics, which in most cases are the principal reasons for their application. An important aspect concerns shape configurations, ranging from very simple to extremely complex, resulting in manufacturing economies that are now being used more frequently for cost reduction. To date, most of the innovation has taken place in commercial products, since those markets took immediate advantage of the performance/cost gains. The shapes that are produced to meet packaging, medical, auto, electronic, toy, leisure, and other markets include mostly complex shapes. They include integral handle lids, snap fits (chapters 19 and 20), and others. Hollow and structural industrial shapes are blow molded into such forms as automotive sun visors, chair seats, and fuel tanks.

Bottle Design

Beverage suppliers with requests for attractive, lightweight, and more hot-fill, single-serve bottles continue to focus on new blow-molded designs. Heat-set PET bottles typically weigh 5% to 10% more than other bottles, owing to extra material used for vacuum panels that are located on their sides. These panels prevent bottles from warping or crumpling from the vacuum created by hot-fill contents; but, these panels have limited PET bottle designs for familiar shapes. Hot-filling remains the prevalent bottling method in North America, so processors are forced to be creative in developing containers that meet performance requirements and offer appealing designs.

An example of this development is from Snapple Beverage Corp. They will replace the glass packaging it uses for its flagship products (ice teas and juice drinks) with 20-oz hot-fillable stretch blow-molded PET bottles from Graham Packaging, York, Pennsylvania. They are equal in appearances, with novel bell-shaped upper profiles, and offer greater appeal to consumers with plastics' lighter weights and resistance to breakage. The bottles feature Graham's Active Cage vacuum panel technology. An Active Cage panel can be smaller in size, thinner, and allows the frame of the bottle walls to move or flex while absorbing the hot-fill vacuum, as opposed to the traditional rigid frame that encases the panel. The Active Cage panel also tightens as the vacuum is absorbed, increasing bottle strength.

A newcomer to the plastic packaging industry is multinational water bottler Ty Nant, Bethania, Wales. It enlisted industrial designer Ross Lovegrove, renowned for Apple's Mac computer. He created what are believed to be the first asymmetric PET bottles having no visible ribs. Ribs for strengthening the bottle walls are concealed within the contours of the package. The bottles, which have ripple effects and are organic shaped, come in different sizes. They are blow molded at a new plant using six-cavity stretch injection blow machinery from ADS, Cergy-Pontoise, France (536).

Industrial Products

Industrial blow molding continues to take advantage of blowing complex parts in one piece, resulting in a number of benefits. Blow molding continues to challenge injection molding, thermoforming, and rotomolding. Industrial blow molding by the automotive industry has led everyone who works in or is involved in plastic development to exploit and expand its market opportunities. There are three basic design approaches in industrial blow molding: hollow, double-wall, and structural shapes. Today, they are used to market a variety of industrial as well as consumer parts. All these design methods offer protection, versatility, and economy for applications with which other processes have difficulty to compete.

Whether one discusses uneconomical or impractical approaches to tool up large structural parts in injection molding, whether we are designing products with limited volumes that need energy-absorbing or insulating properties, or whether we simply want to reduce part weight, the blow-molding process is an excellent choice to consider. Further, in comparison with metal forming or some other plastic processes, such as thermoforming, industrial blow molding offers a definite advantage through the elimination of secondary operations (consolidation of parts, modular approaches).

The development of double-wall structure concepts is directly attributable to the work of Peter T. Schurman. In December 1963, Schurman developed double-wall cases that required a substantial male mold member or "core" instead of just two flat cavities with incidental contours therein. Schurman explained that Illinois Tool Works molded, on a developmental basis, flat step treads for use on Hamilton Cosco kitchen stools. They actually were double-wall parts, about $13 \times 8 \times 1$ in, with a level textured top surface and a ribbed underside. Joining two treads with an integral hinge created the first double-wall product, with the smooth side on the outside cavity and the ribbed side on the inside core. From this exciting beginning, the concept of double-wall cases has been used wherever there was a need for special products, such as those to store and protect high-value instruments and major tools.

In addition, the sales appeal of a permanent storage case developed the market opportunities that corrugated packaging could not offer. Double-wall cases are being used for chain saws, small power tools, monitors for heart pacers, computers, security chests, musical instruments, carrying cases (Fig. 6.115), and many other applications. Possibilities are almost unlimited. HDPE and propylene copolymers, with or without fillers, are the resins of choice, concurrent with new resin developments, such as advanced engineered thermoplastic alloys.

Complex Irregular Shape

With regard to development and use, the products now on the increase are those that have extremely complex 3-D shapes. They are targeted for all markets, particularly irregular hollow-shaped industrial parts. Often a single blow-molded part can replace an assembled part made of other materials, substantially reducing costs. The process adapts to both low- and high-volume production runs. The process is flexible as to part size and shape. Part sizes range from extremely small to very large, with the only restriction being the size of the machine available.

Other important characteristics that relate to the plastic used include low weight, thermal and electrical insulation, corrosion resistance, chemical resistance, permeability resistance, transparency, low coefficient of friction, color, aesthetics, and capability of monolayer and multilayer. In regard

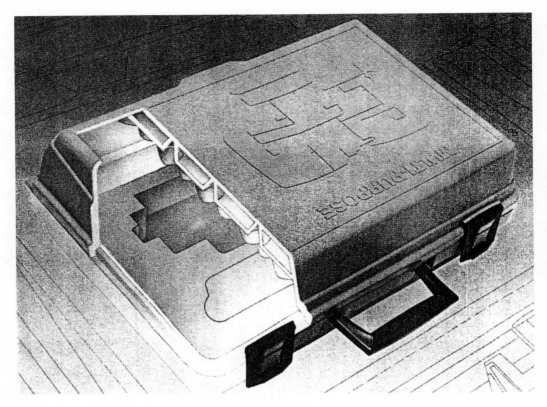

Figure 6.115 Examples of extruded blow-molded double-wall HDPE carrying case, which protects and simplifies part storage

to flammability, hot-enough fires can destroy all plastics, like other materials. Some burn readily, others slowly, others with difficulty, while still others do not support combustion upon removal of the flame (chapters 21 and 22).

Very complex irregularly shaped products have been extending state-of-the-art blow-molding technology, because conventional mold designs fail to produce acceptable wall thicknesses. Using mold halves with moving sections, some complex products are routinely and successfully molded. When the product geometry involves reverse folds or inwardly protruding channels and narrow projections, conventional molds cannot guarantee acceptable wall thicknesses. In plastic-starved areas, particularly corners, product walls are thinner than in other sections and are thus subject to rupture. The solution is in using an alternative to the conventional moving section.

Complex Mold Action Equipment has been developed and applied to moving mold sections to blow mold highly irregular product shapes (similar action is used in other molds, such as injection-molding molds). The significance of the technology is that it provides a means to BM complex products in one piece. Although it is fairly common to move mold sections to release undercuts, which permits product removal, molding techniques that affect or control wall thicknesses are used by relatively few blow molders. The ability of these new molders to blow mold uniquely shaped products has paid dividends by eliminating or reducing costly secondary operations.

ORIENTED 3-D PARISON

The 3-D extrusion blow-molding processes have been reviewed. In conventional extrusion blow molding, the parison enters the mold vertically rather in a straight tube. In 3-D BM, the parison is oriented in an open or closed mold. It is manipulated in the tool cavity, providing complex geometric products that can have uniform or nonuniform wall thicknesses, corrugated and noncorrugated sections, and so on. It provides a means to significantly reduce scrap/flash, waste, and quality (3).

Dr. Michael Thielen and Frank Schuller introduced this process some years ago (528 to 535). In the meantime, a certain number of different systems became established in the market using single and multilayer plastics with hard-soft-hard plastics. Systems permit different materials (single or coextrusion) to be sequentially extruded, alternating one after the other so that a parison with different sections of material in the extrusion direction can be produced. The processes include suction blow molding, 3-D blow molding with parison manipulation, and a split-mold horizontal machine with a vertically opening mold and a six-axis robot laying the parison into the cavity or a machine without a closing unit.

With 3-D, substantial savings are possible, which may be achieved by using blow-molding machines specifically designed for the production without (or at least with significantly reduced) pinch lines. In these processes, the extruded parisons (with diameters smaller than the articles' diameters) are deformed and manipulated then moved directly into the mold cavities so that the remaining lengths of the pinch lines are reduced to minimums. The 3-D manipulation of the parison with programmable manipulators or up to six-axis robots and special devices allows the pinchless production of complex articles.

The process has advantages that include lower clamping force required, less effort for deflashing, very little if any flash, no refinishing work on outer article diameter, improved quality of the article due to wall thickness distribution, and no reduction of strength due to pinch lines.

Molds are comparably expensive; however, the suction blow-molding process requires simple and inexpensive molds. Here, the parison is ejected into the closed mold and "sucked" through the mold via an airstream provided by a blower system. After exiting at the lower end of the mold, the parison is squeezed off by closing devices and the inflation and cooling process can follow.

A simple and inexpensive mold can be installed on a horizontal 3-D machine. This mold is installed in a vertically opening clamp. The lower mold half slides out under the head, where a manipulating robot places the parison into the cavity. After laydown, the mold half slides back under the top mold half. The clamp closes and the inflation and cooling processes follow.

Major problems involved in high-strength/strong-curved conventional blow moldings can produce irregular wall thicknesses at inner and outer radii, a result of the different stretch ratios.

This was solved with a radial wall-thickness distribution system (RWDS). It permits a balancing out of the wall thicknesses between the inner and outer radii of the curved sections, even with parisons that have small diameters. To guarantee a regular wall-thickness profile of the final product, the wall thickness is adjustable not only concerning the length of the parison but also as to its circumference.

OTHER DESIGN APPROACHES

In addition to products, processes, and design approaches reviewed in this book, there are many more. Figures 6.116 and 6.117 provide examples of BM-stacked (two or more) products or parts during a single BM operation followed with manual or automatic cutting action to separate the parts. They can include snap-fit, in-mold labeling, and others.

During extrusion blow molding, molds can be moved and flash removal at the bottleneck using the blow pin is quick and efficient (Fig. 6.118).

To reinforce an injection or extrusion blow-molded container, opening plastic inserts that are injection molded are used (Fig. 6.119).

The shape of an extruded parison can be used to provide for a living hinge (chapter 20) (Fig. 6.120).

There are bellows-collapsible bottles produced in conventional extrusion blow-molding equipment (Patent #4,492,313). These foldable bottles provide advantages and conveniences such as (a) reducing storage, transportation, and disposal space, (b) prolonging product freshness by reducing oxidation and loss of carbon dioxide when contents are removed, and (c) providing continuous surface access to foods, such as mayonnaise and jams (Fig. 6.121) (3).

SUMMARY

The plastics industry is one of the most important business sectors providing significant contributions to the economy and standard of living across all sectors worldwide. In this industry, blow molding is the third major process following extrusion and injection molding. Plastic blow-molded products are used in many different markets, including foodstuff, beverage, packaging, medical, toy and game, recreation, transportation, appliance, marine, agriculture, water filter, missile and rocket, U.S. postal service, toilet and water conservation, and others.

HISTORY

Extensive information is available concerning the history of BM worldwide. Only a few are presented here. The first patent for extruded polymer parison molding was U.S. Patent #237168, filed May 28, 1880, and issued February 1, 1881. The patent was issued to Celluloid Novelty Co. and Celluloid Manufacturing Co., New York. For over the past century, there have been extensive technological and new product developments, with many more developments and innovations occurring now.

The major commercial development in BM started with the commercial availability of low-density polyethylene (LDPE) during World War II. In the 1940s, Monsanto blow molded an LDPE squeeze bottle. The industry started its growth with the liquid bleach and detergent market. This market was rapidly converted and replaced glass, which is bulky and dangerous to health and

Shuttle Machine Limitations

- The key limitation of shuttle machines is that only one or two molds are available to deliver the desired output.
- To raise the output of the machine, the number of parisons and mold cavities must be raised.
- The axial mold shuttle distance increases greatly.
- Parison adjustment and control become more difficult, especially in multi-layer applications.
- The number of blow pins, knives, trimming stations and IML baskets also increases proportionally.

Overcoming Shuttle Machine Limitations

- One possible approach is to "stack" mold cavities vertically, forming bottles in a tail-to-tail approach.
- This would require both top and bottom blow pin calibration and more complex trimming and handling mechanisms.
- Extruding and capturing the longer parison accurately is more difficult.

Figure 6.116 A shuttle EBM machine limitation and solution (courtesy of Graham Plastics Group)

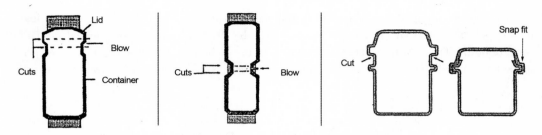

Figure 6.117 Views of multiple action extrusion blow-molding containers

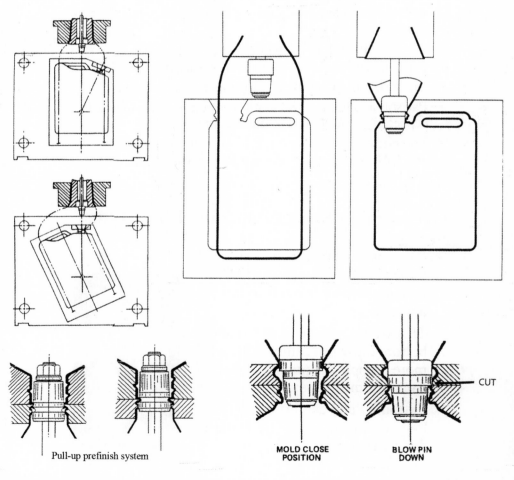

Figure 6.118 Schematics of moving molds and removing bottleneck flash (courtesy of Uniloy Milacron)

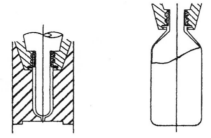

Figure 6.119 Example of inserting a plastic injection-molded reinforcement into a blow mold

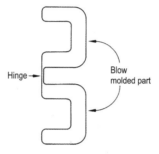

Figure 6.120 Living hinge is part of the extruded blow-molding parison

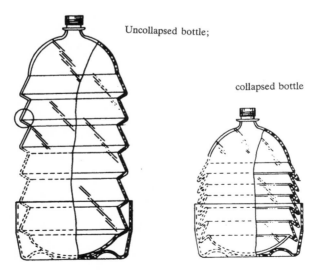

Figure 6.121 Collapsible bottle capable of 85% size reduction or 75% volume reduction

property due to breakage. The industry moved right through household products, such as shampoos and liquid soaps.

The largest potential market in packaging is the food industry, and in the late 1950s with commercial development of high-density polyethylene (HDPE), the milk market emerged with a plastic gallon (3.8-liter) jug that swept the industry, nearly replacing glass and plastic-treated paper. Early in the 1960s, an unprecedented expansion of BM facilities took place accompanied by general public acceptance of plastic bottles. Different types of plastics are now used to produce various BM products that range from simple to complex shapes.

During 1958, the Coca-Cola bottle was injection stretched blow molded using acrylonitrile-styrene (AN) plastic. Unfortunately, after production started in about eight plants with efficient recycling facilities on the East Coast, AN was not permitted to be used by the FDA because of possible food contamination (carcinogenic, etc.), even though its permeability requirements really could not be determined by instrumentation available at that time.

After decades of wasting money and time by the government (taxpayers) and the cost and time used by the plastics industry, AN (acrylonitrile) received an OK. In the meantime, that development was the forerunner in using PET plastics in an avalanche of bottles to appear commercially worldwide that started during the 1970s and continues in high gear. Originally, rumors had it that "competitors" fed FDA the "wrong" information.

During the mid-1950s, Coor's (beer) Company in Colorado almost went into using commercially stretched injection blow-molded bottles. They would have used the injection blow molding with rotation process. Unfortunately, they were using acrylonitrile-styrene plastic (from Barex plastic of Sohio of BP Chemical International; D.V.R. IMM equipment project). As just reviewed with the Coca-Cola bottle, AN could not be used per the FDA.

The ketchup bottle started in 1983 by H. J. Heinz, USA. The popular stretched squeezable coextruded BM bottle used a PP/EVOH barrier and PP plastics with adhesive interlayers. Later, PET/EVOH and other combinations that included interlayers of recycled plastic followed.

History continues to be made worldwide.